Dalton's Introduction to Practical Animal Breeding

Dalton's Introduction to Practical Animal Breeding

Fourth edition

Malcolm B. Willis
BSc (Dunelm), PhD (Edin), Hon. Assoc. RCVS
Senior Lecturer in Animal Breeding and Genetics
Faculty of Agriculture and Biological Sciences
University of Newcastle upon Tyne

**Blackwell
Science**

Blackwell Science Ltd
Editorial Offices:
Osney Mead, Oxford OX2 0EL
25 John Street, London WC1N 2BL
23 Ainslie Place, Edinburgh EH3 6AJ
350 Main Street, Malden
  MA 02148 5018, USA
54 University Street, Carlton
  Victoria 3053, Australia
10, rue Casimir Delavigne
  75006 Paris, France

Other Editorial Offices:

Blackwell Wissenschafts-Verlag GmbH
Kurfürstendamm 57
10707 Berlin, Germany

Blackwell Science KK
MG Kodenmacho Building
7–10 Kodenmacho Nihombashi
Chuo-ku, Tokyo 104, Japan

First edition of *An Introduction to Practical
  Animal Breeding* by D.C. Dalton
  published by Granada Publishing Ltd
  1980
Reprinted 1981
Second edition published by Collins
  Professional and Technical Books 1985
Reprinted by BSP Professional Books 1989
Third edition entitled *Dalton's Introduction
  to Practical Animal Breeding* published
  by Blackwell Science 1991
Reprinted 1993, 1996, 1997
Fourth edition published 1998

Set in 10.5/13.5pt Times
by DP Photosetting, Aylesbury, Bucks
Printed and bound in Great Britain by
MPG Books Ltd, Bodmin, Cornwall

DISTRIBUTORS

Marston Book Services Ltd
PO Box 269
Abingdon
Oxon OX14 4YN
(*Orders:* Tel: 01235 465500
  Fax: 01235 465555)

USA
Blackwell Science, Inc.
Commerce Place
350 Main Street
Malden, MA 02148 5018
(*Orders:* Tel: 800 759 6102
  781 388 8250
Fax: 781 388 8255)

Canada
Login Brothers Book Company
324 Salteaux Crescent
Winnipeg, Manitoba R3J 3T2
(*Orders:* Tel: 204 224-4068)

Australia
Blackwell Science Pty Ltd
54 University Street
Carlton, Victoria 3053
(*Orders:* Tel: 03 9347 0300
  Fax: 03 9347 5001)

A catalogue record for this title
is available from the British Library

ISBN 0-632-04947-2

Library of Congress
Cataloging-in-Publication Data
is available

Dedication

In memory of Alan Wilson
student, colleague and friend,
1949–1977

Contents

Introduction

The term animal breeding refers not so much to the mating, reproduction and rearing of animals, as to the application of the principles of genetics to livestock improvement.

Livestock have been domesticated by man for around 14 000 years, with the dog being probably the first such species followed by sheep, goats, cattle and others. This long association with man does not imply that animal breeding is an ancient practice, although the empirical use of modern principles may well have preceded their scientific definition. There is no doubt that the selection of animals into different forms has been practised over a long period of time. Certain types of dog would appear to have been differentiated into something like their modern forms from biblical times.

The eighteenth-century English farmer Robert Bakewell is regarded as the father of modern animal breeding in laying the foundations of the Shire Horse, Longhorn cattle and Leicester sheep. English Thoroughbred horses had their first 'stud book' published in 1791 and Coates' herdbook for Shorthorn cattle appeared in 1822.

Both of these events predated the Austrian monk Gregor Mendel, whose work with the garden pea was formulated in 1865 and was to prove the basis of modern genetics. His work lay largely undiscovered until about 1900. William Bateson, working with chickens, showed in 1901 that Mendelian principles were applicable to animals as well as plants, and modern animal genetics was off and running.

The Hardy–Weinberg law, the basis of population genetics, was formulated in 1908, and since that time a great many scientists have contributed to our understanding of animal breeding in one form or another. Early biometricians working in the field were Galton and Karl Pearson while Sir Ronald Fisher and Sewell Wright advanced their ideas to lay the foundations of modern biometric techniques. Lush at Iowa, Henderson at Cornell, Robertson at Edinburgh and others such as Van Vleck at Cornell/Minnesota have advanced our understanding of the quantitative and statistical nature of animal breeding, both before and after Watson and Crick had defined the nature of the gene for which they won the Nobel prize in 1962.

Although animal breeding in an empirical sense is quite old it is very young in terms of our scientific understanding of the subject. Most agricultural schools, at whatever level, attempt to teach animal breeding and some principles may also be taught in veterinary schools, although in many cases, alas, not as much as one might wish.

Most people involved with practical breeding of livestock, whether these be farm animals, dogs, cats, guinea pigs or the like, are usually fascinated by the concept of animal breeding, but have only a rough understanding of its principles, especially beyond the Mendelian stage. In contrast, agricultural or veterinary students rarely share this enthusiasm, frequently being put off by the mathematical nature of the subject. Many books have been written seeking to overcome this and to put forward often complex ideas in intelligible form. Falconer's *Introduction to Quantitative Genetics*, first published in 1960, has for 30 years been regarded by many as a standard text for more advanced students to the discipline. Books of a more animal breeding nature and a more introductory level are less obvious. Clive Dalton's *An Introduction to Practical Animal Breeding* (1980) proved very useful for introductory courses at my own university, but since Dr Dalton no longer wishes to continue the revision of that book, this one is offered in its place.

The format follows Dr Dalton's in looking first at the nature of what we wish to do with domestic livestock (Chapter 1). Then follows a brief look at the basic nature of Mendelism (Chapter 2). Population genetics is looked at in Chapters 3 and 4, while Chapters 5 to 7 deal with selection. Chapter 8 looks at breeding systems and the remainder of the book (Chapters 9 to 11) looks at practical situations and the future.

A deliberate attempt has been made to retain the format developed over the previous three editions. It has been a deliberate policy to keep mathematics to a minimum. This may offend purists who feel that proofs should always be given, but this has always been intended as an introductory text and the first introduction many people will have to animal breeding after, possibly, a basic genetics course. Too much mathematics could deter some while the enthusiasts will no doubt eagerly seek to expand their understanding by delving into more advanced texts. In this fourth edition some material has been included on horses and companion animals (dogs/cats). This should make the book more interesting to veterinarians since they tend to be in the front line of companion animal breeding. In addition, there is an increase in Animal Science courses at universities and colleges and many of these include considerable material on the horse, dog and cat, to which end this book may help. A bibliography for further reading is given which, with books cited in the references, should help those seeking to take breeding further.

Malcolm B. Willis

Acknowledgement

I am most grateful for the kindness of Professor Frank Nicholas of the University of Sydney who made extensive suggestions about the third edition. I hope that in this new edition I have managed to correct deficiencies and make useful improvements.

M.B.W.

Chapter 1
Traits in Domestic Animals

Classes of traits

Trait, character or variable are terms applied to the features in which man has an interest with respect to his animals. Trait is the way we tend to refer to a particular feature, while character or variable are the terms usually used in statistical terminology. Traits need to be carefully defined or described but essentially fit into five basic characters:

(1) *Fitness traits* These are not concerned with athletic prowess but rather with features of reproduction and viability. Conception rate, litter size, calving interval, gestation length and survival ability are all examples.

(2) *Production traits* These include features like milk yield, growth rate, feed efficiency and weaning weight. They are concerned with features that cover performance of the animal in productive terms, but must be defined as to the period of time involved. Thus, growth rate must define the time scale, e.g. growth from birth to weaning or from weaning to slaughter.

(3) *Quality traits* Carcass composition, backfat thickness in the pig, eye muscle area in all carcasses, meat quality, milkfat percentage – these are all characters which are increasingly important as pressures mount for more acceptable food in health terms.

(4) *Type traits* These are usually features of a more aesthetic nature where personal preference is important and the breeder selects for such things as coat colour, coat type, udder shape and constructional traits. These features are less crucial to breeders of farm livestock than was once the case, but are still the principal basis of selection in species such as the dog or cat. In farm

species the category would include such aspects as legs and feet as well as teat number in pigs, and these features do have a bearing on animal performance.

(5) *Behavioural traits*   These are increasingly important, insofar as animal welfare is concerned. They include not only obvious traits such as herding ability of sheepdogs or tracking ability in police dogs but also docility and response to stress in most species. The ability of a sheepdog to herd sheep depends, for example, not only on the dog but, in part, upon the instinct of the sheep to flock together. Similarly, livestock protection dogs used with sheep must not only possess the instinct but also need to be raised alongside sheep in order to develop the behaviour patterns.

It is a basic principle of animal breeding that the more traits one seeks to include in a breeding programme, the harder the task will become. It is thus essential to decide which traits are crucial and include these but keep this number to manageable levels. Priorities do, of course, change. In the early days of most livestock breeding great stress tended to be placed upon physical appearance or type traits. As more became known about the inheritance of production and quality, type traits became less important. Some features still command attention but have been reduced in number. This has been accentuated as pedigree breeds became less important in certain species, but physical attributes still do command attention among some breeders and can have economic merit to such persons.

Most traits of importance tend to be *polygenically* controlled; that is they are controlled by many genes, but some, especially features affecting such things as coat colour, tend to be relatively simple Mendelian traits with only one or two major genes being implicated for each feature.

The importance placed upon particular traits should depend upon their economic value. To this end production traits are generally of greatest importance as well as being among the easiest to evaluate. In all species reproductive traits have to be important since, without good reproduction, the species will have difficulties with regard to its survival. Litter-bearing species have great emphasis placed upon litter size or survival to weaning.

Measurement of traits

Traits can be classified as being either subjective or objective. Most live animal traits can be measured, e.g. weight gain, milk yield, fleece weight, litter size. Some are subjectively assessed using a hedonic scale as in the case of type evaluations and some carcass attributes. Some measurable traits are not easy to use or define as to their value. For example, assessing racehorses can be done on the time they take to run specific distances or upon prize money won or upon Timeform ratings (based on handicaps and weight carried). These are all relatively easy to do but difficult to interpret as to their significance and meaning. If a horse is well ahead of the field its jockey may slow it down once winning is inevitable, but this means that speed achieved is not maximised. In assessing a racehorse we are, of course, not always able to divorce its performance from that of the jockey on its back.

Breeding is frequently about making comparisons between one animal and another, or more particularly the traits measured in these animals. It is thus important that measurements are made on valid terms. It is, for example, difficult and unwise to compare animals of differing ages or to compare those reared and managed in different herds or flocks.

Adjustment or correction factors have to be applied before some data can be legitimately analysed and comparisons made. Comparison within herds and years as well as age groups, such as is often undertaken with dairy records, is one way of eliminating herd/year/ season and age effects, but this is not always feasible.

Some of the factors which need to be considered before comparisons can be made include:

(1) *Herd/year/season or year/location within the country* might all be lumped together as management factors. Often these can only be reduced by making comparisons within specific groupings. Thus, if animals born in the same herd in the same year and the same month are considered as contemporaries and assessments are made within this grouping, these effects are largely eliminated.

(2) *Age of animal* Weaning weights of suckled calves will vary because calves tend to be weaned at a specific point in time but

are born over a period of several months. Comparing weights of animals at different ages is invalid so corrections need to be made to a constant age.

(3) *Sex of the animal*   This may be more important in some species than others (e.g. cattle more than sheep), but much depends upon the point in time. Males tend to be bigger and to grow faster and perhaps be leaner than females, so corrections are needed or the sexes should be assessed separately.

(4) *Litter size*   Individuals in large litters may grow less quickly than those in smaller litters, so adjustment to a constant litter size is called for, or the use of covariance, a statistical technique for analysis which takes out the effects of litter size variation, is needed.

(5) *Dam age*   Young dams may produce slower growing progeny than older dams because they have lower milk production. Thus, correction for dam age may be required.

(6) *Parity*   This refers to birth order, e.g. first litter, second litter, etc., and although related to dam age the two traits are not necessarily identical. Parity must therefore be considered in analyses. Heifers, for example, are much more likely to experience calving difficulties (dystocia) than mature cows.

Correction factors usually have to be calculated from the data being assessed though standardised factors are sometimes used. The validity of standardised factors is open to debate but need not concern us here. In general terms correction factors are either additive or multiplicative.

In correcting the weaning weight of females to that of males, it is important to know not only the mean weights for each sex but also the standard deviations. If males are heavier than females but the standard deviations are more or less identical for the two sexes then additive factors can be used. A fixed amount is added to each female weight to make it the equivalent of what it might have been had it been a male. If, however, the females have not only a lower weight but also a smaller standard deviation, then they should be corrected by multiplying their weight by a given factor which not only raises the weight to the male equivalent but also enlarges the variation.

Traits of importance

Traits vary from species to species; they also vary in importance at different periods of time and to different segments of the agricultural industry. If a milk producer derives income largely from milk yield then total milk production will be more important than milk composition. If payment depends in part upon milkfat and/or protein content, then these assume greater importance to the animal breeder. The suckler calf producer selling calves at weaning will have less interest in post-weaning gain than will the farmer buying those calves. If the national sheep industry is one based primarily upon meat production (as in Britain) then wool represents a relatively small part of the breeder's aims. This is in sharp contrast with Australia where wool is a principal source of income and selection has been directed towards that product rather than towards meat. A pig breeder is interested in litter size and litters produced per year, but the buyer of weaned pigs has no interest in this, being more concerned with post-weaning gain.

Tables 1.1 to 1.4 show some of the economically important traits in different farm livestock. Not all will be equally important to all countries or to all areas of the industry, but all are of importance to some. The heritabilities (see Chapter 5) are given only in broad terms because this feature is largely something peculiar to the data from which it was derived and thus detailed values are of less importance but appear later (Table 5.1).

In general, low means less than 15%, medium runs from about 20 to 50% and high refers to values above 50%, though high values rarely exceed 70%. High heritabilities indicate characters that are capable of showing faster progress, while low indicates characters that may not respond well to selection.

The lists presented in Tables 1.1 to 1.4, while fairly comprehensive, are not intended to be either exhaustive or in any order of priority. Indeed, the priorities may vary from breeder to breeder as well as from species to species. Priorities will depend, in part, upon economic considerations. Many scientists may argue, for example, that type characteristics should have a low priority, but many pedigree breeders would stress that animals of good physical appearance as regards their relationship to the breed ideal or breed standard can command premium prices and thus type has economic merit for

Table 1.1 Economically important traits of dairy cattle

Trait	Heritability
Reproductive	
age at first calving	low
services per conception	very low
service period	very low
calving interval	low
twinning	low
dystocia	low
Yield per 305-day lactation (or per year or lifetime)	
milk	moderate
milkfat	moderate
protein	moderate
solids-not-fat	moderate
Composition (percentage basis)	
milkfat	high
protein	high
solids-not-fat	high
Other features	
somatic cell count (as indicator of mastitis)	moderate
aspects of type	range from low to high (mostly moderate)
rate of milk let-down	low to moderate
temperament	low

them. Others might argue that type characteristics are, in some instances, related positively to certain production traits, although this is more readily checked by scientific study.

A breeder of Suffolk sheep selecting for physical attributes of head and leg shape may be losing out on important traits such as growth rate and lean meat production, but if he can sell his rams for many thousands of pounds to other breeders with similar interests there is clearly economic gain to be made.

Priorities may differ according to country and to location within a country. For example, in Britain a farmer of hill sheep will have fewer lambs per ewe mated than a lowland sheep farmer. Wool will thus be a greater proportion of a hill farmer's income as well as being more

Table 1.2 Economically important traits of beef cattle

Trait	Heritability
Reproduction	
age at puberty/first mating	low
number of live births	low
dystocia	low
services per conception	low
Maternal ability	
weight of calf weaned per cow	moderate
mothering ability	low
temperament	low
Weight	
birth	moderate
weaning (approx. 200 days)	moderate
yearling	moderate
pre-weaning gain	moderate
post-weaning gain (feedlot)	high
400-day weight	high
Carcass traits	
killing out (dressing) per cent	moderate
fat depth (various locations)	moderate
longissimus dorsi area	high
classification/grade	moderate
weight of saleable meat	moderate
weight of excess fat	moderate
weight of bone	moderate
proportions of lean/fat/bone	low to moderate
ratio meat to bone	low to moderate
ratio first to second quality meat	low to moderate
muscling/shape	moderate to high
eating quality (tenderness, etc.)	low to moderate
Other traits	
type aspects (various)	low to high (mostly moderate)
disease resistance	low
heat resistance (in tropics)	low to moderate

Table 1.3   Economically important traits of pigs

Trait	Heritability
Reproduction	
litter size	low
litters per year	low
number of piglets born	low
survival ability (of piglets)	low
Weight	
weaning	low to moderate
slaughter	moderate
pre-weaning gain	low to moderate
gain weaning to slaughter	moderate
age at slaughter	moderate
Carcass	
carcass lean	moderate
killing out (dressing) per cent	moderate
fat depth (various)	moderate
longissimus dorsi area	high
carcass length	moderate
quality traits	low to moderate
halothane	single gene
Others	
type (rarely important)	low to high
teat number	moderate
temperament	low
skin colour	simple inheritance

crucial to the survival of lambs in the harsh hill environment. Accordingly, emphasis upon fleece weight and type will differ from the hill to the lowland situation.

In a species such as the dog the priorities of a police dog breeding establishment using German Shepherd dogs or the Guide Dogs for the Blind Association, breeding mainly Labrador and Golden retrievers, will be different from those of breeders seeking to produce companion animals. In the same way, a Labrador breeder using his or her dogs for shooting activities will have different priorities from those of the breeder who exhibits in the showring. These differing priorities can lead, and have led, to differences in mental and physical

Table 1.4   Economically important traits of sheep

Trait	Heritability
Reproduction	
number born	low
age at first mating	low
litters per year	low
Maternal ability	
milk production	low to moderate
temperament	low
survival ability of lamb	low
Weight	
birth	low to moderate
weaning	low to moderate
slaughter	moderate
pre-weaning gain	low
post-weaning gain	moderate to high
age at slaughter	moderate
Fleece	
weight	moderate
fibre diameter	moderate
staple length	moderate
follicle parameters	
ratio primary to secondary	moderate
medullation (hairiness)	moderate
colour	moderate
Carcass	
killing out (dressing) per cent	moderate
classification/grade	moderate
fat depth	moderate
proportions (meat/fat/bone)	low to moderate
longissimus dorsi area	high
Others	
type and breed characteristics	low to high (usually moderate)

attributes within the same breed of dog. The same is true of Thoroughbred breeders producing sprinters or long-distance horses and those whose horses go over hurdles or fences.

Chapter 2
Mendelian Principles and Laws

The cell

All individuals of whatever animal species are made up of millions of minute cells. A typical cell is illustrated in Figure 2.1. Each cell contains a nucleus in which are found the *chromosomes*. These thread-like structures are constant in number within a given species though anomalies can sometimes occur as regards this number. Chromosomes come in pairs (called *homologous* pairs) with one member of each pair coming from each parent. The numbers of chromosomes found in particular animal species are shown in Table 2.1. The term *diploid* refers to the total number of chromosomes found in the animal's cells while *haploid* refers to the half number or number of pairs. The haploid number is found in the *germ cells* (eggs and sperm).

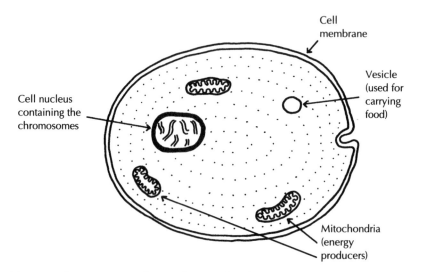

Fig. 2.1 The main parts of a simple animal cell

10

Table 2.1 Chromosomes in various species

Species	Chromosome number	
	Diploid	Haploid
Man	46	23
Cat	38	19
Pig	38	19
Rabbit	44	22
Sheep	54	27
Goat	60	30
Cattle	60	30
Bison	60	30
Donkey	62	31
Horse	64	32
Dog	78	39

Body cells have the ability to duplicate themselves by what is termed somatic growth which involves a process called *mitosis*. The body cells of, say, a cow will contain 30 pairs or 60 chromosomes. During mitosis the chromosomes duplicate themselves into what are called *chromatids* which are still joined at a point called the centromere. There are various stages of mitosis, which need not concern us here, but at some point the chromatids separate and go to each end of the cell such that both members of each pair are at each end. A constriction develops in the centre of the cell which then forms two cells, each of which has the full 60 chromosomes of the original cell. Mitosis is thus a duplicating process.

Germ cells are formed by a process termed *meiosis*. In this there is chromatid formation, but at one stage of the process there is a reduction division such that each new cell is made up of only half the chromosomes. However, there is one member of each homologous pair in each cell so that the egg or the sperm carry (in the case of cattle) 30 chromosomes. When egg and sperm meet up, the new *zygote* will once again contain 60 chromosomes in 30 homologous pairs. Although the proviso exists that each fertilised egg must contain the 30 pairs, there is an element of randomness in terms of which chromosomes end up in which haploid cell. On average one would expect that an egg cell contains 15 chromosomes from the cow's mother and 15 from the cow's father, but this is by no means certain, nor are the chromosomes allocated in any set order. This means that there is a considerable element of chance about which genes an individual inherits. Figure 2.2 shows a very simplified illustration of mitosis and meiosis.

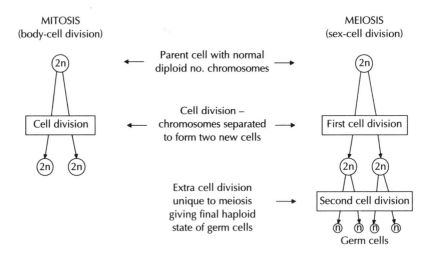

Fig. 2.2   Mitosis and meiosis contrasted

Genes

Genes used to be likened to beads on a string, with the genes being the beads and the string the chromosome. In reality genes are an integral part of the chromosome so the analogy is not correct. Deoxyribonucleic acid (DNA) is the principal component of the chromosome and it was in 1953 that the structure of this was first discovered by Watson and Crick who, with a colleague, Wilkins, shared the 1962 Nobel prize. The DNA molecule is a two-stranded affair known as a double helix. The strands are made up of nucleotides which contain deoxyribose (a sugar), a phosphate group and a nitrogenous base. This latter is made up of adenine (A), guanine (G), thymine (T) and cytosine (C). The A and G bases are called purines and the T and C are pyrimidines. It appears that A always pairs with T and C always with G.

The sequence of bases is important in determining what is called the genetic code. The bases in particular sequences of three form *codons* which in turn stand for particular amino acids. The sequence GAG stands for glutamic acid and a single change to GTG would convert this to valine. The simple change of one base in a sequence can alter the whole effect of the gene.

A specific gene is normally always found at a particular point (*locus*) on a specific chromosome. Because chromosomes are paired there are two such loci and thus every animal has two versions of a gene. For many loci the two versions are identical and the animal is said to be *homozygous* for that locus (gene). However, many genes have alternatives, caused by a different base order at that locus. If an animal has two different versions at a locus then it is said to be *heterozygous*. Different versions of a gene are called *alleles*. A gene may have several alleles, but an individual animal can only have two versions of any allele because it only has the two homologous chromosomes.

The assignment of genes to specific chromosomes and to specific locations on that chromosome is called *gene mapping* and in the last decade or so such mapping is being undertaken on various species including man, the pig, the dog, cattle and sheep with varying degrees of success. This has enabled certain genes, for example the halothane gene in pig breeds and the progressive retinal atrophy (PRA) gene in the Irish setter, to be located. Now, screening of animals can be used to identify those which carry specific genes, even in a single dose. These mapping projects tend to be international with geneticists in various countries collaborating.

Dominants and recessives

Let us take the gene which caused black or brown/red pigment in a species like cattle or the dog. There are two alleles, one of which causes black pigment to form and the other which causes brown pigment. The black version is 'stronger' because the presence of one of each allele will result in the animal having black pigment not brown. We therefore call black the *dominant* allele and brown the *recessive*. Traditionally, dominant alleles are given uppercase letters and recessives lowercase. The black/brown gene has the letters B = black and b = brown/red. Because there are two alleles in each animal there are three alternative combinations: BB, Bb and bb.

Animals with the genetic make-up BB or Bb will normally be black and those with bb will be brown or red. What we see (black/brown colour) is termed the *phenotype* while the underlying genetic make-up is termed the *genotype*.

In 1866 Gregor Mendel, a monk working in a monastery garden with peas, published a study on several characters found in peas which was the first modern explanation of genetic inheritance and the mode of inheritance. His work lay largely unread until it was rediscovered in 1900 and formed the basis of what is now termed Mendelian genetics. Mendel argued that the unit of inheritance was what we now call the gene and that genes remained discrete and that characteristics remained unchanged after crossing. The segregation of a gene with two alleles is shown in Figure 2.3.

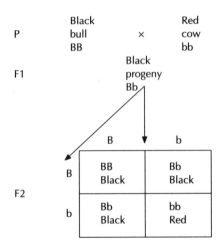

Fig. 2.3   Segregation of gene with two alleles. P is the parent generation and F1 and F2 the first and second filial generations

For a recessive allele to manifest itself in an individual it must be present in duplicate and thus must be carried by each parent. Many deleterious traits tend to be inherited as this kind of simple recessive and when they do arise it can be shown that both parents must be carrying this allele, at least once. Frequently, defects arise from what appear to be two normal parents. Thus, red (and white) Friesians are sometimes born to two black (and white) parents. The red animal must be bb and both parents must be Bb. The white markings stem from a different gene and are inherited independently of the main coat colour.

When crossing a homozygous dominant with a homozygous recessive all the progeny show the dominant trait but are heterozygotes. When these in turn are crossed, dominants and recessives appear in a 3:1 ratio respectively, but the true ratio is 1:2:1 with homozygous dominant, heterozygous dominant and recessive in that order.

If we had used the example of Shorthorns with red or white coat colour and R for red and R^1 for white then the F1 (filial) generation would be RR^1 but would be roan rather than red. Roan is a mixture of red and white hairs but is not a blending of genes. In the F2 generation the ratio would be 1 red: 2 roan: 1 white. Note that these ratios apply over large numbers of matings and would not necessarily apply in small numbers.

Mendel showed that independent genes segregated independently. Figure 2.4 shows the situation when mating a black polled Angus (both of which alleles are dominant) to a red horned Hereford (both of which alleles are recessive). The F1 generation show only the black polled characteristics, but in the F2 generation four phenotypes occur. A ratio of 9:3:3:1 is found (over large numbers) wherein nine show both dominant features, three show one dominant feature, three show the other dominant feature and one shows both recessive features.

If three genes were segregating then a 27:9:9:9:3:3:3:1 ratio would occur with 27 showing all the dominant features, three lots of nine showing each of two dominant features, three lots of three each showing one dominant feature and one showing all three recessive features.

Multiple alleles

As has been mentioned previously (see 'Genes'), more than two alternatives can exist at any locus. In the ABO blood group of man there are three alleles A, B and O and six different genotypes AA, AO, AB, BB, BO and OO. The different groups will form enzymes which lead to antigen formation. Whenever a foreign substance enters an animal it induces an *immune* response wherein the body produces *antibodies* to attack the invader. Antigens are the antibody-

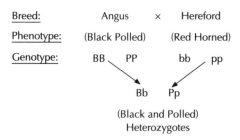

When these heterozygous animals are crossed the results are these:

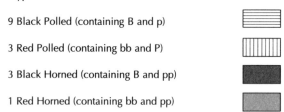

The phenotype ratio from this is:

9 Black Polled (containing B and p)

3 Red Polled (containing bb and P)

3 Black Horned (containing B and pp)

1 Red Horned (containing bb and pp)

Fig. 2.4   Segregation of two independent genes

generating agents. The different blood groups give rise to different antigens. However, AA and AO are indistinguishable as are BB and BO. Thus, although we have six genotypes there are only four phenotypes, A, B, AB and O.

Some genes can have very many alternative alleles and often they are given suffixes. Thus, in the agouti coat colour series found in dog breeds there are A, a^y, a^w, a^s, a^t and a which lead to a variety of coat colours. Not all breeds have all these alternatives, but all breeds carry the gene even if they only have one version of it.

Sex inheritance

Thus far we have been dealing with the concept of paired or homologous chromosomes. One set is not necessarily paired. This set is called the sex chromosome. Female animals have a fairly large identical pair of chromosomes whereas in males there is one large one, akin to those in females, plus a small one. The female pair are termed X chromosomes (females are thus XX) and the small male chromosome is called Y such that males are XY. In birds the sexes are the other way round and to avoid confusion females are usually called ZW while males are ZZ.

Following the meiosis process, eggs carry only one of these sex chromosomes and are thus all X-bearing. In contrast sperm will carry either an X or a Y. If a Y-bearing sperm fertilises the egg then the resultant offspring is male, whereas fertilisation by an X-bearing sperm results in a female offspring.

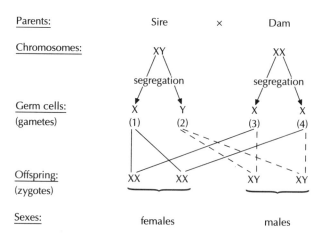

Fig. 2.5   Determination of sex

Although Y-bearing and X-bearing sperm are produced in more or less equal numbers, there is either a predilection of the egg for Y-bearing or Y-bearing are more motile. In any event more XY zygotes are produced than XX. A slightly higher mortality is known among male embryos, but in most mammalian species slightly more males are produced than females. Individual sires may produce abnormally high or low proportions of sons, but in general terms most mammalian species give rise to a 52:48 male to female ratio in every 100 births. *Sex ratio* is the term used to define the number of males per 100 females. Thus, 106 would indicate 106 males per 100 females.

Sex chromosomes carry few genes of importance, but the genes for haemophilia A and haemophilia B as well as Duchenne muscular dystrophy are carried on the X chromosome along with a small number of others. These genes are thus termed *X-linked* genes. Few genes are believed to be carried on the Y chromosome so that the term *sex-linked* genes usually refers to those that are actually X-linked. Chromosomes other than the sex chromosomes are termed *autosomes* and the term autosomal recessive would refer to a gene carried on a non-sex chromosome.

Although extra chromosomes can occur in some individuals this is usually a lethal status. Exceptions would be the extra chromosome 21 in man which leads to Down's syndrome. Extra chromosomes usually result from some error during meiosis. In some species additional sex chromosomes are seen at fairly rare intervals. These are well established in man where one can have individuals which are XXY, XYY and XXX, for example. It is known that in XX cases, only one X chromosome is actually 'switched on' in any cell, though not the same one in all cells. This may account for the non-lethality of some XXY or XXX cases. Abnormal sex chromosome numbers are seen in dog breeds like the Cocker spaniel but are very rare. Tortoiseshell (a mixture of black and orange coat colour) is a sex-linked trait in cats and such animals should always be females. Tortoiseshell males are seen on rare occasions. Often they are XXY and sterile but sometimes they are carrying mixtures of XX and XY cells.

Lethal genes

Some genes act in a way that causes the individual to die either before or subsequent to birth. In the case of a defect that is lethal in the embryo there is likely to be an abnormal segregation ratio. If the aa version was lethal and only AA and Aa animals survived there may be a 1:2 ratio instead of a 1:2:1. A lethal gene may be one in which there is a single change in a base but which leads to some serious biochemical change in the animal. Some genes act by causing the build-up of products that would normally be excreted. These are termed storage diseases. It is well established that dropsy (bulldog calves) is lethal in Ayrshire cattle though rarer now than it once was. Hairlessness, a dominant gene, is lethal in Chinese Crested dogs when present in a homozygous state as it may well be in most animal species. Von Willebrand's disease (a factor VIII blood disorder) is believed to be lethal in a homozygous state, as is dominant white coat colour in horses. Amputate in Friesian cattle, A-46 (under-development of the thymus) in Black Pied Danish cattle, and imperforate anus in pigs and sheep are all lethal recessives.

Some lethals like Huntington's disease in man are dominants which eventually cause death but at 40 plus years rather than early in life. Some defects do not cause death but bring about serious imperfec-tions which are termed semi-lethals. In livestock most badly deformed animals would be culled even if they did not succumb to the condition they had inherited so that semi-lethal may underestimate the severity of the problem.

Linkage

An exception to the Mendelian law of independent segregation concerns linkage. This occurs when particular genes are at different loci but carried on the same chromosome. Because chromosomes tend to be passed on in their entirety, different genes are transmitted together. Chromosomes do, however, tend to crossover with others and parts may become interchanged so that a chromosome may be transmitted minus a part of it but instead transmit that part from its homologous partner or from another chromosome altogether. The chance of such 'breakages' obviously is greater the further apart two genes are from one another. Genes which are close together on a particular chromosome may be transmitted together more often than those which are further apart. Linkage may affect segregation ratios but can be valuable in detailing chromosome mapping. Linkage may inhibit new combinations but it can enable existing ones to be retained.

Sex-linked, sex-limited and sex-influenced traits

Sex-linkage has already been mentioned in connection with sex chromosomes (see 'Sex inheritance'). Haemophilia A is one such trait carried on the X chromosome. If we designate the normal clotting allele as H then the recessive allele which affects the coagulation factor (factor VIII) is termed h. Females can be HH, Hh or hh, though this last genotype is rarely met with. In contrast males are H0 or h0 where 0 represents the Y chromosome devoid of this gene. Males that are h0 are affected by haemophilia even though they have only one recessive allele. Such a male would pass his X chromosome to all his daughters who would thus carry h and *all* be carriers. In contrast his sons would all be normal, assuming he was mated to HH females. Any female that is Hh is a carrier and will pass the h allele to half of her offspring, giving rise to males of which half are affected and to daughters of which half are carriers. To obtain affected females would require the use of affected males and carrier females which would be a rare event.

Sex-limited traits are those which are expressed by only one sex. Milk yield and egg production are obvious sex-limited traits since males cannot express these features, though they do carry the genes which influence them. In contrast cryptorchidism is a male trait that cannot be seen in females. Genes for these traits are carried by both sexes even though expressed in only one sex. This is why progeny testing of dairy bulls and cockerels is important.

Sex-influenced traits occur in both sexes though there will be a preponderance in one sex. Some examples of horns in sheep are sex influenced since the same genotype will produce polling in females but horns in males. Mahogany red coat colour in Ayrshire cattle is similarly more likely in males. By the same token hip dysplasia, which is a condition commonly seen in the dog and rarely in man, is more frequent in females than males.

Epistasis

Although different genes segregate independently (linkage excepted), it is possible for a gene at one locus to influence the expression of a gene at a totally different locus. This is termed *epistasis*. It differs from dominance in the sense that dominance refers to alleles of the same gene while epistasis refers to different genes. However, both are involved with gene interaction.

Coat colours in various dog breeds are often influenced by interaction between genes. An example of this is coat colour in Labrador retrievers. There are three phenotypes: black, yellow and chocolate (liver). To be black the animal must carry the dominant allele B at least once, but for this colour to be expressed the animal must carry at least one copy of the dominant gene E from the extension series. To be chocolate a dog must carry the recessive combination bb but again requires the E allele at least once. Yellow colour is caused by the presence of the recessive combination ee which is a dilution feature and which prevents black pigment from forming and thus leads to a yellow coat colour. However, ee does not affect nose colour so dilute blacks (B-ee) have black noses. In contrast bb affects nose colour causing it to be brown or liver. All bb dogs have brown/liver noses and dilute liver dogs (bbee) have such noses. There are nine different genotypes resulting from the interplay of these two genes but only four genotypes. Only one genotype (bbee) can be identified directly from its phenotype. Black dogs, of which there are four kinds even though all may appear identical, can range from the BBEE animal whose progeny will all be black regardless of which Labradors it mates through to the BbEe combination which could give rise to all four phenotypes.

Table 2.2   Epistatic effect on coat colour in Labradors

Phenotype	Genotypes			
Black	BBEE	BbEE	BBEe	BbEe
Chocolate	bbEE	bbEe		
Yellow (black nose)	BBee	Bbee		
Yellow (liver nose)	bbee			

Chapter 3
Principles of Population Genetics

Economic traits

In Chapter 1 various traits of possible economic importance in farm livestock were listed (see 'Traits of importance'). Most of these traits are complex not only in terms of the factors which influence them but also in terms of the genes which contribute to their value. The traits dealt with in Chapter 2 (see 'Sex-linked, sex-limited and sex-influenced traits') tended to be those controlled by single genes, but most important features of livestock are controlled by many genes. The term polygenic has already been mentioned (see Chapter 1, 'Classes of traits') and refers to traits which are influenced by several genes each of which may be relatively minor but which, collectively, have a cumulative importance. In addition, environmental factors play their part. The coat colour of an animal will not, usually, be affected by the way that animal is fed and managed. However, the milk yield of a cow or the growth rate of a pig will be influenced by numerous environmental factors. This is illustrated in Figure 3.1 which shows some of the factors that might influence a trait such as 400-day weight in beef cattle.

Variation

Anyone dealing with biological data realises that there is variation in whatever characteristic one examines. Some traits show minimal variation, e.g. calf numbers born to beef cows will range over 0, 1 or 2 and only rarely exceed the level of twins, though 3, 4 and 5 calves are known to occur. In contrast, the trait of 400-day weight will vary over quite a large range.

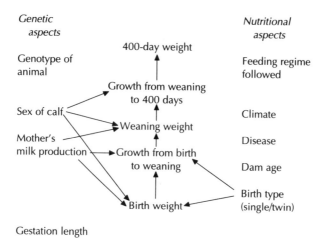

Fig. 3.1   Some factors affecting 400-day weight in beef cattle

For purposes of illustration, assume that 400-day weight is controlled by three pairs of genes each with two alleles A/a, B/b and C/c. Also assume that the basic combination aabbcc gives a 400-day weight of 350 kg and that each uppercase letter (whether A, B or C) adds 20 kg to the basic weight. Further assume that environmental influences are non-existent. An individual of genotype AABBCC will weigh 470 kg compared with the aabbcc animal weighing 350 kg. Consider a mating between these two. The resultant progeny will be of genotype AaBbCc and will thus weigh 410 kg which is midway between the two parental weights of 470 and 350 kg. Now mate individuals of genotype AaBbCc. These represent the F1 generation and will give rise to the F2 generation. The A/a, B/b and C/c alleles will segregate independently and the F2 generation will appear as shown in Table 3.1.

The 64 individuals seen in the F2 generation exhibit all the weights (in 20-kg units) from the original parental low of 350 kg to the original parental high of 470 kg. The frequencies are shown in Table 3.2. This illustrates considerably more variation in the F2 compared with the F1, but the overall mean is the same in both the F1 and F2 generations. It will also be seen that although there are 64 individuals they divide into 27 distinct genotypes.

The 27 different genotypes in Table 3.2 actually show segregation into seven different phenotypic categories. Expressed another way,

Table 3.1 Segregation of three genes affecting 400-day weight (kg) in beef cattle (hypothetical)

Male gametes	*Female gametes*							
	ABC	ABc	AbC	aBC	Abc	aBc	abC	abc
ABC	AAB BCC 470	AAB BCc 450	AAB bCC 450	AaB BCC 450	AAB bCc 430	AaB BCc 430	AaB bCC 430	AaB bCc 410
ABc	AAB BCc 450	AAB Bcc 430	AAB bCc 430	AaB BCc 430	AAB bcc 410	AaB Bcc 410	AaB bCc 410	AaB bcc 390
AbC	AAB bCC 450	AAB bCc 430	AAb bCC 430	AaB bCC 430	AAb bCc 410	AaB bCc 410	Aab bCC 410	Aab bCc 390
aBC	AaB BCC 450	AaB BCc 430	AaB bCC 430	aaB BCC 430	AaB bCc 410	aaB BCc 410	aaB bCC 410	aaB bCc 390
Abc	AAB bCc 430	AAB bcc 410	AAb bCc 410	AaB bCc 410	AAb bcc 390	AaB bcc 390	AaB bCc 390	Aab bcc 370
aBc	AaB BCc 430	AaB Bcc 410	AaB bCc 410	aaB BCc 410	AaB bcc 390	aaB Bcc 390	aaB bCc 390	aaB bcc 370
abC	AaB bCC 430	AaB bCc 410	Aab bCC 410	aaB bCC 410	Aab bCc 390	aaB bCc 390	aab bCC 390	aab bCc 370
abc	AaB bCc 410	AaB bcc 390	Aab bCc 390	aaB bCc 390	Aab bcc 370	aaB bcc 370	aab bCc 370	aab bcc 350

there are 27 individuals showing all three uppercase letters (ABC). There are nine individuals in each of the AB, AC and BC categories. There are three individuals each of which exhibit A, B or C and one which is aabbcc.

This is the exact distribution mentioned in Chapter 2 (see 'Dominants and recessives') for a three gene trait. However, in this example

Table 3.2   Gene combinations for 400-day weight (hypothetical)

Gene combination	Number	400-day weight (kg)
AABBCC	1	470
AABBCc	2	450
AABbCC	2	450
AaBBCC	2	450
AABBcc	1	430
AAbbCC	1	430
aaBBCC	1	430
AABbCc	4	430
AaBBCc	4	430
AaBbCC	4	430
AABbcc	2	410
AAbbCc	2	410
AaBBcc	2	410
AaBbCc	8	410
AabbCC	2	410
aaBbCC	2	410
aaBBCc	2	410
AabbCc	4	390
AaBbcc	4	390
aaBbCc	4	390
aabbCC	1	390
aaBBcc	1	390
AAbbcc	1	390
aabbCc	2	370
aaBbcc	2	370
Aabbcc	2	370
aabbcc	1	350

it was assumed that A, B and C had exactly the same effect. We can reassess the data merely by looking at upper case and lowercase letters regardless of the actual letter and Table 3.2 can then be produced as a histogram showing simply the number of uppercase letters present and the weight that ensues. This is shown in Figure 3.2.

Although the divisions in Figure 3.2 are quite broad (20 kg), the pattern exhibited is one in which there are few animals at the extremes and more towards the centre. This is a typical pattern seen

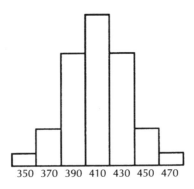

Fig. 3.2   Histogram of 400-day weights in Table 3.2

in polygenic traits and if the top points of the histogram are joined in a continuous line the pattern approximates to what is called a Normal curve.

The 400-day weight example is an oversimplification in that many more than three gene pairs are involved in the genetic determination of the character. Moreover, no account was taken of sex differences nor of any environmental influences. The fact that environmental influences act upon the genetic make-up to modify the phenotype means that the smooth balanced pattern of the histogram would be modified. There would, in reality, be much smaller divisions than 20 kg and possibly a large spread of weights though this would depend upon breed. The maximum weight used here would be fairly average for many British-based breeds (see MLC, 1995). The example, while hypothetical, illustrates something about variation which is a fundamental feature of polygenically controlled traits.

Animal breeders are interested in seeking uniformity and there are many practical advantages in having animals perform in similar fashion. The nature of genetics, plus environmental influences, makes this impossible and a geneticist would regard variation as highly desirable if selection is to be effective.

One way to illustrate variation is to show the distribution of a particular trait in diagrammatic form. Figure 3.3 shows the birth weights of single-born Scottish Blackface sheep out of 2-year-old mothers. Fairly broad divisions have been used and the balance each side of the centre which was seen in the hypothetical 400-day weight data is partly lost. Nevertheless, these real data show the same normal distribution pattern expected in most traits.

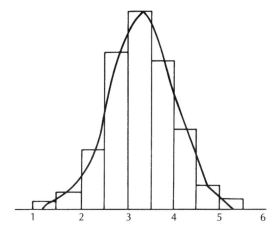

Fig. 3.3 Birth weights (kg) of 1290 Scottish Blackface lambs

Features of Normal curves

Normal curves, regardless of the trait for which they are drawn or the units of measurement, show certain features which always apply if the trait is truly normal. The curve is bell-shaped with the highest point around the mean and decreasing numbers as the extremes are reached. Curves can vary considerably, such that they could be very spread out along a large base or narrow with a high peak and a small base.

In dealing with biological data the assumption is often made that normality is present. If it is not, data may have to be transformed (e.g. to logarithms) before analyses can be undertaken. Populations have variation which in statistical terms is called *variance* and designated by the symbol σ^2. If each individual is called X and the number of individuals is n then the variance of a population is given by the formula:

$$\sigma^2 = \frac{\sum X^2 - \sum \frac{(X)^2}{n}}{n - 1}$$

A high variance would indicate a very variable character and a low variance a higher peak type of curve with a small spread along the baseline. However, variance is expressed in terms of units squared. It measures the squared deviations of each item from the overall mean. Kilograms squared is not a very meaningful term and thus a better measure of variance is the standard deviation or SD. This is the square root of the variance and is thus expressed in actual units of measurement.

In any Normal population one SD either side of the mean would encompass about 68% of the population. Two SDs would encompass about 95% and three SDs would cover about 99%. Thus, as a rough guide, there will be about three SDs each side of the mean or about six SDs in total range. If we had a breed with a 400-day weight of bulls averaging 500 kg and the SD was 30 kg then the breed would range in weight from about 410 to 590 kg. Just over two-thirds of the animals would range from 470 to 530 kg and some 95% from 440 to 560 kg. A bull weighing 680 kg is six SDs above the mean and would be so extreme that, in all probability, it is from a different population (breed) to the original group.

Most important traits in farm livestock vary along the lines of a Normal curve and such characters are often called *continuous variables*. Most characters which are polygenic in their mode of inheritance follow these patterns but not all. Litter size in pigs, cats or dogs would follow this pattern, but that would not apply to cattle, sheep or horses. Figure 3.4 shows a more typical pattern for sheep. It is skewed, with a long tail towards the upper end. Application of Normal curve statistics to such data would be inappropriate and transformation to logs or square roots would make such data more normal and thus allow analyses to be validly made.

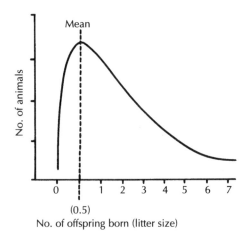

Fig. 3.4   A skewed distribution (litter size in sheep)

Traits such as litter size in sheep are called discrete variables and different statistical methods apply for their analysis. The same may be true for populations which are preselected according to specific limits. Figure 3.5 illustrates a population from a herd book wherein entry into the book required the cow to give a specific minimal yield of milk. The curve is a truncated Normal curve. Yields of less than 5000 litres are excluded. Application of Normal curve statistics to such a population would be incorrect. It is a prerequisite for studies of genetic parameters that the population being used is a normal one and not one that is biased in any way.

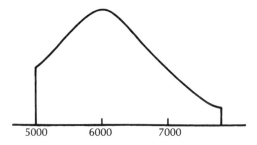

Fig. 3.5   Truncated curve of milk yield (litres)

Animal breeders are working, in the main, with populations which display normal variation and in which the object is to bring about improvement. There may be considerable debate about which characters to improve and why, but the usual object is to improve the mean value of the trait. Improvement may, of course, not necessarily be in an upward direction. Reducing dystocia in most species or backfat thickness in the pig would be examples of improvement that reduced the mean but in doing so would improve the trait. In essence the animal breeder is seeking to change the gene frequencies in the population by increasing the frequency of desirable genes and reducing the frequency of undesirable ones. The problem is that, apart from simple Mendelian traits, breeders do not know how many genes are involved; still less are they able to identify them. They are, therefore, working in a kind of empirical fashion. This does not mean that animal breeding is a hit and miss affair, although there is clearly some element of chance (luck) involved. Although most genes cannot be identified, traits can be measured or evaluated and certain standard techniques can be used to bring about progress.

Many advances in the development of breeds and types of livestock were undertaken long before Mendel and it is often claimed that 'practical' breeders were better than 'scientists'. Practical breeders tended to work by certain rules of thumb for which genetic support is sometimes forthcoming. For example, the old adage 'breed the best to the best' meant breed the best males to the best females, with 'best' referring to excellence in some specific feature, usually appearance. This is fairly sound genetic practice for polygenic traits although it will not always apply.

Threshold traits

Although most polygenic characters show a normal curve type of distribution, there are characters which exhibit an 'all or nothing' type of appearance which may appear akin to a Mendelian trait but which are controlled by many genes. Survival of young is one such trait where there are only two alternatives (survival or death) but where the underlying genetic pattern may be complex. Cryptorchidism is another feature with the added complication that it can be seen only in males. Twinning in cattle could be considered as an all or nothing trait in that the cow either does or does not give birth to twins.

Although such traits may exhibit discrete patterns as opposed to the pattern of a Normal curve it is known that some of these characters are controlled by several genes. Suppose, for example, that cryptorchidism has five genes influencing it, each with two alternatives (0 or +). In the threshold situation there will be some point at which there are sufficient plus (+) alleles to induce a cryptorchid state. Suppose that this point is seven plus alleles. All individuals with fewer than seven plus alleles would appear normal with testicular descent into the scrotum. Those with seven to ten plus alleles would exhibit retained testicles. The point at which the phenotypic change will occur is termed the threshold point and, thus, such traits are termed *threshold traits*.

Some threshold characters can have several divisions. Thus, patent ductus arteriosus (a heart disease) occurs when the ductus arteriosus (a fetal condition) persists after birth. Normally this duct closes postnatally, but if it only partially closes it is termed ductus diverticulum

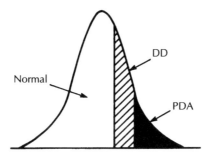

Fig. 3.6   A threshold pattern. DD, ductus diverticulum; PDA, patent or persistent ductus arteriosus

(DD) whereas if it remains open it is patent or persistent ductus arteriosus (PDA). Figure 3.6 illustrates this.

Threshold traits are particularly difficult, both as a concept and in terms of seeking to influence them. If only both normal and abnormal versions exist, then both are recognisable and culling of the abnormals may be easy. However, one cannot readily distinguish between those normals which are very close to the threshold point and hence potentially 'dangerous' and those normals which are far removed from the threshold point and thus are unlikely to give rise to abnormal offspring.

Sometimes all or nothing traits can be classified into a series of categories or grades as opposed to merely two or three. This aids the analysis and management of the problem. Hip dysplasia, a defect seen frequently in many breeds of dog (as well as in cattle, horses, rabbits, cats and man) can be classified as normal or affected. It can also be assessed in multiple grades (7 to 10) or on a scoring scheme as is done in the UK and Australasia using a scale from 0 to 53 per hip (see Willis, 1989). Scoring changes the assessment of the trait from an all or nothing feature to something approaching a normal curve albeit with a skew towards the higher end. Ways of selecting threshold traits and assessing genetic parameters are given in Chapter 5.

Chapter 4

Ways of Changing Gene Frequencies

Introduction

Animal breeders are interested in 'improving' their animals, but there may be considerable disagreement about what constitutes improvement. Some breeders will seek to alter production traits, others will be looking for improved physical appearance, yet others for combinations of these. The features they are examining are complex characters which are controlled partly by the animals' genes and partly by environmental features. What breeders are actually doing is altering the frequencies of particular genes, even if they may not know which genes they are dealing with. A breeder seeking to improve weaning weight in beef cattle may be weighing cattle at weaning and selecting the heaviest. In part, selection will be for animals which produce more growth hormone per unit of body weight and thus will increase the frequency of those genes and lower the frequency of those alternative alleles which do the opposite. To understand the principles it is useful to look at the single gene situation which involves awareness of the Hardy–Weinberg law.

The Hardy–Weinberg law

The Hardy–Weinberg law was first documented in 1908 by Hardy, an Englishman, and Weinberg, a German, working independently of each other. Essentially it applies to large populations and its principal statement is that there will be no change of gene frequencies provided that there is:

(1) Random mating (each animal has equal opportunity to mate any animal of the opposite sex).
(2) No mutation.
(3) No migration (animals do not enter the population from outside it).
(4) No selection.

If the gene frequencies do not change then the population is said to be in equilibrium. The essential features of this are the large population so that chance fluctuations are avoided or are negligible, together with the random nature of mating and the absence of any changing forces (mutation/migration).

It is not within the scope of this introductory book to go into the proofs of the Hardy–Weinberg law, but understanding of the principles involved has a bearing upon the understanding of how to change populations in the desired direction.

Assume a random mating large population of Friesian cattle with no forces seeking to change frequencies (no mutation or migration). Assume that red coat colour (bb) appears in 1% of the population. For purposes of this example the white markings (caused by genes at a different locus) are ignored. At this black/red locus we have two alleles: B = black and b = red. Because red cattle appear we have three genotypes BB, Bb and bb but only two phenotypes black and red. The popular convention is to designate the B allele as p and the b allele as q. Since there are only two alleles, p + q must equal 1.0. The Hardy–Weinberg law states that in the conditions wherein it applies (as shown above) there will be p^2 of the BB genotype, q^2 of the bb genotype and 2pq of the heterozygote Bb genotype. If there are 1% red animals then that is equivalent to q^2. If we express this as a proportion then bb $= q^2 = 0.01$. This means that its square root is q and equals 0.10. If q = 0.10 then p = 0.90 so that together p and q equal 1.0.

In a Hardy–Weinberg equilibrium situation we can calculate that the frequencies of the three genotypes are: BB = 0.81, Bb = 0.18 and bb = 0.01. Because black is a dominant allele we cannot distinguish between the BB and Bb animals since all are black. We can only say that 99% are black and 1% red but 18% of the 99% black carry the red allele.

Assessing deleterious genes

The Hardy–Weinberg law is often used when seeking to assess the frequency of undesirable alleles in a population. Suppose that the frequency of odematous (dropsical) calves in an Ayrshire cattle population was one animal per 300 calvings, what would be the incidence of carriers in the population? Since dropsy is an autosomal recessive, one calf in 300 births represents a frequency of 0.0033 which is equivalent to q^2. This means that $q = 0.0577$ and thus $p = 1 - q = 0.9423$. The proportion of carrier animals would be $2pq$ or $2 \times 0.9243 \times 0.0577 = 0.1067$. This means that about one in every ten animals is a carrier of dropsy. This was the probable situation in the Ayrshire breed in Britain back in the 1950s, but the incidence of both the disease (as well as the breed) has since declined.

Factors changing Hardy–Weinberg frequencies

Mutation

A change in a gene probably caused by some mistake during the replication of DNA alteration will occur from time to time but may not be of any consequence because it may be corrected or not passed on. However, such a change which is transmitted is known as a mutation. A mutation can occur as a one-off, in which case it may be quickly lost unless it has some selectional advantage. Some mutations may be recurrent and occur at regular intervals with a given frequency.

Mutation rates tend to be very low, ranging from 10^{-4} to 10^{-8} for most cases. Such rare events will tend to be of minimal significance, but over a long time period could have a slight effect upon frequencies. If allele A mutates to a at the rate of u and allele a mutates to A at a rate of v then when $pu = qv$ there will be a point of equilibrium. If the mutation is in one direction only then the mutating allele will gradually decline in frequency. Because mutational rates are so slow their effect upon Hardy–Weinberg tends to be minimal.

Migration

Migration is the introduction of breeding animals from one population into another. A great deal of genetic change results from this activity and migration is thus much more important than mutation. The policy of grading-up, wherein sires of a second breed are used in successive generations in an existing breed, is an example of migration. The virtual demise of the Shorthorn in Britain during the mid-1950s and beyond did not occur simply by culling the breed from herds but rather by the successive use of British Friesian sires in a grading-up policy which reduced the Shorthorn to the stage of a rare breed. In turn, since the mid-1960s, the British Friesian has found itself being 'graded-up' to the Holstein-Friesian by the increasing use of Holstein sires, mainly from North America. A similar policy has applied to other breeds in different species and locations.

The effect of migration upon Hardy–Weinberg frequencies will depend upon the proportion of migrants used and the difference between their gene frequencies and those of the native population. This is given by the formula:

$$\Delta q = m \ (q_m - q_o)$$

where m is the proportion of migrants used and q_o and q_m are the gene frequencies in the original and migrant populations respectively. The change of frequency is indicated by Δq.

Assume a cattle population that is all horned and thus all pp, into which polled bulls (PP) are introduced to undertake 15% of the mating. The original frequency of the P allele in the original population (q_o) was 0 and that of the migrants (q_m) was 1.0. The value for m is 0.15 and since polled bulls replace horned ones the formula for change becomes:

$$q_1 = \frac{15}{2} \ (1.0 - 0) = +0.075$$

The division of 15 by 2 occurs because only males are introduced not females. In one generation the frequency of the polled gene has increased from 0 to 0.075 or by 7.5%. Note that 15% of the cattle would be polled but they would all be Pp in genotype so the gene frequency is only half of that figure. Further use of polled bulls in future generations would further increase the frequency of the P allele.

Selection

Whether artificial or natural, selection can have a marked effect upon gene frequency, although not necessarily as marked as some forms of migration. However, migration is effective only up to a given point, whereas selection can be an ongoing feature. When considering selection at the single gene level, it is important to specify the type of gene action. Selection may be against a recessive allele or against a dominant allele or it could be in favour of a heterozygote or there could be sex-linkage to consider. It is beyond the scope or intent of this book to go into great detail on these possibilities, which are effectively dealt with in several books listed in the Further Reading section. It does, however, seem appropriate to look at selection against a recessive allele, because in some livestock species practical breeders are concerned with trying to eradicate deleterious genes that are often recessive.

In selecting for or against a gene, one has to look at the phenotypes and their relative 'fitness'. In this context fitness refers to the capacity of the particular phenotype to survive and reproduce taking account of breeder choice. In most domestic livestock species it is man who makes the decisions about which animals are to be retained for breeding and this is termed *artificial selection*. It does not always follow that these plans come to fruition. Animals selected as potential breeding stock may die or prove infertile which means that *natural selection* still has some influence upon breeding actions. Fitness thus refers to those phenotypes which the breeder thinks are the most suitable for reproduction in the context of what the breeder is trying to achieve.

In the example chosen of a dominant allele with selection against the recessive version there are three genotypes AA, Aa and aa. Because AA and Aa are phenotypically indistinguishable they are, in practice, given equal importance, whereas aa is selected against. The theory is laid out in Table 4.1. A fitness of 1 is given to the desired phenotypes (A) though not all may survive to reproduce. The undesired phenotype is selected against to the value of s (the coefficient of selection). If aa was lethal then s would equal 1 and all

Table 4.1 Selection against a recessive allele

Phenotype	AA	Aa	aa	Total
Frequency	p^2	$2pq$	q^2	1
Fitness	1	1	$1-s$	
Gametic contribution	p^2	$2pq$	$(1-s)q^2$	$1-sq^2$

aa animals would naturally be culled. An s of 30 would indicate that for every 100 animals of the favoured phenotype used, 70 $(100-30)$ of the unfavoured would be used.

Although the original frequency totalled 1 the gametic contribution becomes $1-sq^2$ because some of the aa individuals are culled from breeding. The original frequency of the A allele was made up of all the AA and half the Aa animals. This is still true in the generation after selection but now the population total is no longer unity. The frequency of the A allele in the next generation (p) thus becomes:

$$p_1 = \frac{p_0}{1 - sq_0^2}$$

The frequency of the a allele then becomes $1 - p_1$.

Consider that the red allele (b) in the Holstein exists at the level of 1% of calves born (a frequency of b of 0.10) and suppose no red animals are bred from, such that $s = 1$. The situation then is as shown in Table 4.2. The frequency of p (B) is 0.90 and that of b is 0.10 but substituting in the formula above gives:

$$p = \frac{0.90}{1 - 1(0.10)^2} = 0.909$$

The frequency of the red allele (q) will be $1 - 0.909 = 0.091$ which means that red calves will appear at the rate of 0.091^2 or 0.0083. The

Table 4.2 Selection against red coat colour

Phenotype	Black	Black	Red
Genotype	BB	Bb	bb
Frequency	0.81	0.18	0.01
Fitness	1	1	0

original generation produced 1% red calves which has now declined to 0.83% after one generation of selection. In effect, selection against the recessive was not very effective. This is not because of any lack of severity in the selection programme; after all, culling every red is making it as severe as a lethal. Rather it is because the original incidence of red calves was quite low.

It is easier to select against alleles that have a high frequency than against those that have a low frequency. Moreover, as selection against a rare allele continues it becomes harder to reduce it further because most of the undesirable alleles are carried by heterozygote (Bb) animals.

In species like the dog or cat great store is laid on culling defects, but the reduction of incidence of a very rare defect is laborious and not always a fruitful occupation. It can be shown that to reduce a gene from a frequency of q_0 to q_t:

$$\frac{1}{q_t} - \frac{1}{q_0} \text{ generations}$$

Consider the dropsy gene in Ayrshire cattle with a frequency of one calf in every 300 born and a desire to halve this incidence to 1 in 600 calves by not using any dropsical animal for breeding. In the vast majority of cases dropsical cattle are not viable anyway and tend to die early. The original frequency is 0.0033 (see 'Assessing deleterious genes') or a q value of 0.577 while the intended frequency is 0.0017 or a q value of 0.0408. This would take 7.2 generations to achieve. If the generation interval (see Chapter 5) in Ayrshires was about 6 years then the task would take 43 years to achieve. Despite all the effort dropsy would not have disappeared but simply declined in incidence. Of course, greater reduction would ensue if known (proven) carriers of the dropsy allele were culled.

Detecting carriers of deleterious recessives

As shown earlier (see Chapter 2, 'Dominants and recessives'), most species carry deleterious genes. In man over 1500 simple genes exist which are deleterious in their mode of expression and it is not improbable that similar numbers exist in other species, although less attention is given to such features in farm livestock than in man or species like the cat and dog. Nevertheless, breeders often wish to know if particular animals carry such defects which are generally, though not exclusively, simple autosomal recessives. Various techniques exist to determine carriers of such defects and these are discussed in turn. The same principles will apply to other recessive alleles which may not be deleterious, such as those affecting coat colour which may be of importance to breeders of dogs, cats and horses though less crucial in farm livestock.

Pedigree information

Known or likely carriers of recessive alleles can sometimes be predicted from pedigree data. If the animal being studied shows the dominant allele (e.g. N) then the 'worst' that it can be is Nn but it may be NN. If one parent was nn then it is obvious that the animal must be Nn since that particular parent can only have given n to its offspring. Similarly if both parents had been known to previously produce nn progeny but were themselves N, then it follows that these parents must both be Nn and the apparently normal offspring in question has a 2:1 chance that it is NN. Simple Mendelian ratios coupled with chi-square tests can be used to estimate ratios and chances from pedigree data. However, in many cases pedigree data may be totally absent and other techniques are required.

Test mating

Test mating is usually regarded as the mating of a suspect 'carrier' (N?) to either a known carrier (Nn) or an affected individual (nn). The most reliable method is to use nn individuals for testing, but this is only feasible if the nn genotype is viable (as it would be with a coat colour allele but not with some serious physical anomaly). If the N? male is mated to an nn female and an affected offspring results then the sire is proven to be Nn. However, failure to produce an nn case does not exonerate the N? individual. An estimate of risk thus needs to be made for the results of various matings.

Test mating of this kind is only useful if the affected parent can be bred from and the defect is fairly early in onset. There is little point in test mating a sire for a defect that may not appear in the offspring until they are middle-aged since by that time the sire being tested may well be deceased or no longer fertile.

In cases where the nn female cannot be used the alternative is to use carrier females (Nn). They need to have been so proven from pedigree data or previous progeny. The risk of producing nn offspring is less in the N? × Nn mating than the N? × nn mating so that more progeny are needed by this technique to 'exonerate' a sire to a given level of probability.

Two other methods of test mating would be to mate the N? sire to daughters of a known proven carrier (Nn) or to mate the suspect sire to his own daughters. Both have equal probabilities in terms of the identification of carriers. The differences between the two methods are therefore mainly practical ones. The former technique is clearly quicker than the latter. However, mating a sire to his own daughters not only will test him for the problem under study but also may throw up other recessive problems, hitherto not suspected.

Table 4.3 shows the relative probabilities of being wrong if one assumes that the sire is NN for each of the four methods described. If in any single case an nn offspring results then the N? sire being tested is Nn. However, suppose a sire is mated to an nn female and six normal offspring result. From Table 4.3 the chance of being wrong is 1.56%. This means that if we tested 200 sires on such a mating and on six progeny we would be wrong 1.56% of the time in declaring the sire to be NN and in about three cases sires that were actually Nn would slip through the net. The probability of detecting a sire as a carrier is the reciprocal of the figures in the table.

Thus, if we mate our N? sire to nn mates and have six progeny there is 98.44% chance of detection. When assessing relative risks by mating to nn or Nn individuals the chances assessed are additive. Thus, a litter of eight would be the same as eight litters of one off-spring. In the case of matings to daughters the values in the table should be used as proportions and multiplied together. Therefore, a sire mated to five of his own daughters and producing litter sizes of 3, 4, 4, 5 and 6 (all of which progeny were normal) would have a chance of error of:

$$0.71 \times 0.66 \times 0.66 \times 0.62 \times 0.59 \times 100 = 11.3\%$$

Therefore, after quite extensive matings and considerable time there

Table 4.3 Test mating errors (%) when testing for a simple autosomal recessive. Percentage chances of being wrong when assuming N? to be NN because no nn progeny resulted

Normal progeny seen	Mating to affected female (nn)	Mating to carrier female (Nn)	Mating to daughters of proven carrier or sire's own daughters
1	50.0	75.0	88
2	25.0	56.3	78
3	12.5	42.2	71
4	6.25	31.6	66
5	3.13	23.7	62
6	1.56	17.8	59
7	0.78	13.3	57
8	0.39	10.0	55
9	0.20	7.51	54
10	0.10	5.63	53
11	0.05	4.22	52
12	0.02	3.17	52
13	0.01	2.38	51
14		1.78	51
15		1.34	51
16		1.00	51

is still a relatively high chance that if we declare the sire to be NN we have an 11.3% chance of being wrong.

Random mating

One of the problems with test mating as outlined above is that either known affected or carrier individuals have to be maintained for testing purposes which, with deleterious recessives, may be costly or very impracticable. Moreover, one is making matings which while they may result in seemingly normal animals *could be* or *are* producing carrier individuals. One is therefore faced with the problem of what to do with individuals born from such test matings which could be Nn in genotype and thus undesirable for mating purposes. In addition, testing would have to be done for each defect separately. Frequently advocated in species such as the dog or cat – where at least there are litters to benefit from – there are many disadvantages. Test mating progeny may slip back into the breeding population, thereby undoing the good work of testing in the first place. The prevention of this by putting down test mating progeny would find little favour from welfare groups or breeders reluctant to kill ostensibly normal animals. Although test mating Irish setters for progressive retinal atrophy (PRA, a serious eye disease) was practised in the 1940–50 period in Britain with some success it was achieved at great cost in animals and considerable heartache to breeders.

There is great merit in breed societies or controlling bodies having some kind of recording scheme to evaluate matings in the population at large. If the incidence of a defect is known for the population, then the number of progeny needed to 'exonerate' a sire mated at random can be assessed. At the same time as testing for a specific defect, other defects may also be recorded if the recording scheme is supported by breeders. The probability of detection can be shown as equal to:

$$1 - (1 - 0.5q)^n$$

which means that the chance of detection depends upon the frequency of the defect and the number of progeny. If the defect is lethal, and thus there are no females available for mating that are aa, then the probability of detection is reduced to:

$$1 - \frac{(1 + 0.5q)}{(1 + q)^n}$$

For different genotype frequencies in the female population and for different degrees of detection, the numbers of progeny needed in each case are given in Table 4.4. For relatively rare alleles the number of progeny required could be quite large although not difficult to achieve with some artificial insemination (AI) bulls which can produce thousands of progeny. Even some dog sires in popular breeds would be able to amass several thousand progeny, but very rare defects would be detected only fortuitously. It can, of course, be argued that if a defect is very rare then it presents no real problem to the population/breed under study.

Although many consider it desirable to minimise deleterious recessives and other such defects, means to combat them must be kept in perspective. Breeding is about trying to produce desirable

Table 4.4 Numbers of progeny needed to detect carrier status with random mating in the population (95% probability)

Frequency of a allele in population	aa females in population	aa females not in population
0.50	11	17
0.30	18	24
0.10	58	64
0.01	598	604

animals (whether this be efficient producers or merely those of desirable conformation) and such objectives should not be placed in jeopardy by plans aimed at reducing rare defects, the effect of which on the population at large is, by definition, minimal even if very serious for the animal afflicted.

Gene mapping

Mapping the genomes of animals will increasingly allow identification of harmful genes, even if that may not be the principal objective of the exercise. The so-called halothane gene in pigs and PRA in the Irish setter are just two examples of genes which have been identified, and animals can thus be screened using a small sample of blood or other tissue. This will allow not only affected (nn) but also carrier (Nn) animals to be detected without the expensive and time-consuming efforts of test mating. The incidence of PRA in the Irish setter has proved to be comparatively low, doubtless as a result of the test mating done 40–50 years ago and it would be possible, if desired, to screen all breeding stock and eliminate nn and Nn cases from the breeding programme. Since nn cases will go totally blind they could be euthanised humanely at an early age. It is interesting that PRA is found in several breeds, but, as yet, they have not been found to possess the same gene in the same place as the Irish setter. Exactly how quickly genome mapping will locate and identify genes and how many it will identify is unclear, but it could lead to many advances in animal breeding within the next decade that were undreamt of a few decades ago.

Chapter 5
Selection for Polygenic Traits

Introduction

Thus far, emphasis has been mainly on simple traits controlled by a single gene locus. Most economically important traits of farm livestock are not only controlled by many genes (polygenic) but also influenced to a greater or lesser degree by environmental factors. In seeking to improve a character such as weaning weight of beef cattle there is (as yet) no way of knowing how many genes are involved, still less the genotype of individual animals. In order to bring about genetic progress in such a character there is a need to understand certain basic parameters of the population under study.

The assumption is made that the population follows the pattern of Normal curves which allows the properties of such curves to be applied to the data. If data are not Normal then transformation is usually needed. The animal breeder is not seeking to alter the Normality of the trait but rather to alter its mean in either an upward or downward direction. The data on the variable under study would consist of a series of measurements on individuals from the population. The mean, variance and standard deviation can be calculated, and given such data and given populations which have certain relationships, such as siblings and parent/offspring, certain parameters can be calculated.

Heritability

The heritability of a trait expresses that part of the superiority of parents which, on average, is passed on to the offspring. Heritability can be expressed as a proportion or a percentage. It ranges from 0 to 100% or from 0 to 1.0. The symbol h^2 is used to signify heritability.

The phenotypic variance seen in a population is caused by a mixture of genetic and environmental variation. If σ^2 is used to signify variance then the subscripts P, G or E indicate phenotypic, genetic and environmental elements respectively. Thus:

$$\sigma_P^2 = \sigma_G^2 + \sigma_E^2$$

Genetic variation can be subdivided into three components: additive (A), dominant (D) and epistatic (I), while environmental variation can be divided into general or permanent effects (E_g) and specific or common effects, i.e. common to members of a particular litter or to individuals from the same dam (E_s). This means that the above equation becomes:

$$\sigma_P^2 = \sigma_A^2 + \sigma_D^2 + \sigma_I^2 + \sigma_{Eg}^2 + \sigma_{Es}^2$$

The additive part is of the greatest interest since heritability is defined as:

$$h^2 = \frac{\sigma_A^2}{\sigma_P^2}$$

Heritability is a ratio so that decreases in any other sources of variation will increase the heritability. Heritabilities are thus specific to particular populations although there is tendency to find similar values for the same characteristic in different populations. In broad terms heritabilities were given in Tables 1.1 to 1.4 and in Table 5.1 some more specific figures are given, albeit not necessarily applicable to all populations. In most cases a range is given to encompass figures seen in published literature.

Calculating heritabilities

Various techniques exist to estimate heritabilities using either regression or correlation techniques. If data for particular traits (e.g. growth rates) exist for both parents and offspring, then regressions of offspring on mid-parent (the average value for sire and dam) can be calculated. The regression coefficient would equal the heritability. Some traits (e.g. milk yield, egg production) are not measurable in sires so that mid-parent analyses are obviously not possible. However, data on one parent, usually sires (albeit not in these two

Table 5.1 Examples of heritabilities in livestock (%)

Cattle		Sheep		Pigs	
Calving interval	0–15	Lambs born	0–15	Litter size	0–10
Calves born	0–15	Lambs weaned	0–10	Pigs weaned	0– 7
Milk yield	30–40	Weaning weight	10–40	Weaning weight	0– 8
Fat yield	25–45	Fleece weight	30–40	Daily gain	21–40
Fat %	32–87	Staple length	30–60	Feed efficiency	20–48
SNF %	53–83	Fibre diameter	40–70	Killing out	26–40
Protein %	48–88	Crimps/cm	35–50	Backfat C	62–65
Lactose %	28–62	Medullation	34–80	Backfat K	42–73
Feed efficiency	40	Birth coat	59–80	Carcass length	40–87
Birth weight	38	Face cover	36–56	Eye muscle	35–49
Weaning weight	20–50	Wrinkle	20–50	Leg length	46–50
Eye muscle area	40–70			Fillet weight	31–54
Daily gain	40				
Carcass lean	39				

SNF, solids not fat

examples) because they have more offspring, allows the regression of offspring on sire. In this instance the regression coefficient has to be doubled to give the heritability. Regressions of offspring on dam may be less valid because dams and offspring show similarities caused by maternal influences and these could inflate heritabilities.

Maternal effects are environmental influences, largely nutritional, which progeny have in common with their litter mates as a consequence of being carried in the same womb and then being suckled by the same dam. This means that siblings not only have genetic similarities, one with another, but also environmental similarities.

Most studies use half-sibling (half-sib) correlations wherein the study is based on the progeny of different sires and where data are needed on the progeny but need not be collected on the sire. Usually these half-sib correlations are preferable to offspring/parent regressions because half-sibs may be measured at the same point in time and in a similar (or the same) environment. In contrast, parents are measured at a different point in time from their offspring and may have been assessed in different environmental conditions. However, a half-sib correlation is quadrupled to give the heritability so that estimation errors are also quadrupled. Paternal half-sib correlations are preferred to maternal ones because it is usually easier to get the numbers and because maternal effects are eliminated.

Heritability studies, like most genetic analyses, benefit from numbers. In broad terms, the more sires that are involved and the more progeny, the better the chance of meaningful data, other things being equal. A minimum of ten sires with at least five progeny apiece is required and ideally more. It is better to have more sires than more progeny per sire once minimal numbers of progeny have been reached.

Full-sibs are rarely used for heritability calculations because maternal effects confound the data and the same is true of twin data. Identical twins are not easy to obtain, but if used, the differences between them reflect environmental differences. Heritabilities based upon twins are usually much higher than those based on sibling data and as a result twin- and full-sib data are not of much practical value.

Methods of calculating heritabilities using examples are excellently described by Becker (1984).

A difficulty arises with threshold traits (see Chapter 3). In a single threshold situation animals either have or have not got the trait in question, but the underlying genotype is unclear. Heritability assessment is not as simple as with a normal trait. Suppose that cryptorchidism in a male pig population was 5.0% and it was known that full-sibs of affected pigs had an incidence of 15.0%, then the heritability of the trait could be estimated. The formula required is:

$$h^2 = t/r$$

where r is the degree of relationship and t is the correlation of liability to the trait in the two groups concerned. To calculate relationships we need to know the communality of genes in relatives. Parents and offspring, as with full siblings, have half their genes in common and their relationship is thus $\frac{1}{2}$ or 0.5. Grandparents and grandchildren have $\frac{1}{4}$ of their genes in common and thus have a relationship of $\frac{1}{4}$ or 0.25, which also applies to uncles, aunts, nephews and nieces, while first cousins have a relationship of $\frac{1}{8}$ or 0.125. These relationships are usually known as first, second or third degree respectively.

We also need to know something about truncated Normal distributions. An extensive table on this appears in Falconer and Mackay (1996) of which a much reduced version is shown in Table 5.2. This

Table 5.2   Deviation of the threshold point from the mean in a large population (value in standard deviations)*

Population affected (%)	X	Population affected (%)	X
0.50	2.576	8.0	1.405
0.75	2.432	10.0	1.282
1.0	2.236	15.0	1.036
2.0	2.054	20.0	0.842
3.0	1.881	30.0	0.524
4.0	1.751	40.0	0.253
5.0	1.645	50.0	0.000
6.0	1.555		

* A more extensive table is given by Falconer and Mackay (1996)

table shows the percentage of the population affected by a particular threshold feature. The value X shows the point where this percentage would start, expressed as standard deviations from the mean and assuming a Normal curve. In Table 5.3 the mean superiority of a specific percentage of the population is given, again expressed as standard deviations from the mean. The term t is calculated from the formula:

$$t = \frac{X_P - X_R}{i}$$

where X_P and X_R are read from Table 5.2 for the population (P) and the relatives (R) respectively and i represents the value in Table 5.3 for the population.

Returning to the pig example given above, an incidence of 5% gives an X_P value of 1.645 while that of 15% gives an X_R value of 1.036 (both taken from Table 5.2). The value of i, read from Table 5.3 for 5%, is 2.063. Substituting these values we get:

$$t = \frac{1.645 - 1.036}{2.063} = 0.295$$

Because we were dealing with full-sibs for the R group the relationship is 0.5 so the heritability becomes:

$$h^2 = \frac{0.295}{0.5} = 0.59$$

This heritability is a rough guide and makes the assumption that the variance is the same in both the related and affected animals. This may not always be accurate and the result may slightly underestimate the true figure.

Table 5.3 Selection intensity (infinite population size)

Percentage selected	Mean superiority (in standard deviations)
1	2.665
2	2.421
3	2.268
4	2.154
5	2.063
10	1.755
15	1.554
20	1.400
30	1.159
40	0.966
50	0.798
60	0.644
70	0.497
80	0.350
90	0.195

Selection differential

In seeking to make progress it is important to know not only how a character is inherited but also what sort of selection may be made in terms of degree. Breeders clearly make a conscious decision as to which animals will be used for breeding depending upon the criteria being considered. In theory, a breeder is trying to select the 'best' or most appropriate animals as the parents of the next generation and obvious differences may exist over such criteria. Even if there is unanimity about what is the 'best' these selected animals may not live to reproduce or may prove less fertile than was hoped.

Selection differential (S) is defined as the superiority of the animals selected to be parents over the mean of the population from which they came. The greater the selection differential, the greater the potential progress, other things being equal. Selection differential is largely determined by the breeder when selection is being made, but there are certain factors which limit or control what the breeder can do.

Usually, selection is more effective in males than females. This arises because in almost all species fewer males are required than females and as a result greater intensity of selection can be practised in males, especially if AI is used. An AI bull, for example, could mate 20 000 cows in a year whereas in natural service he may only mate 40 to 50 cows. Thus, with AI, even fewer males may be required than with natural service, allowing still greater selection differentials.

There are physical limitations. Let us assume a beef herd of 100 cows and that cows survive for an average of six calvings. Not all cows will do that but for those culled early others will produce longer. On the assumption that half the calves born are females these 100 cows will, in their lifetime, produce 300 daughters between them. If the herd is to remain at 100 then the breeder must retain 100 of the 300 calves as replacements, assuming there are no outside purchases. The breeder will, hopefully, seek to retain the best 100 calves based on the criterion that is being used. If this was weaning weight, for example, the best 100 weaning weights would be picked. The breeder is thus restricted to selecting the best 33% of his female calves. In contrast, a beef bull used for a couple of years in natural service might give rise to 50 sons so that the best 1 in 50 could be chosen. The best 2% of bulls would be more extreme in merit terms than the best 33% of cows.

Obviously, the breeder will not select at the end of a period but on an annual basis and the breeder may introduce outside material from other herds, not simply select his own. Nevertheless, the example is intended to show how restricted female selection can be compared with that of males. Breeders are selecting animals on merit to be the parents of the next generation. Because the breeder needs fewer males he can be more restrictive in what he keeps on the male side. The breeder is looking at four routes of selection:

- Males to produce males
- Females to produce males

- Males to produce females
- Females to produce females

In dairy cattle it has been estimated that about 75% of progress has resulted from the first two routes and only some 5% from the females to females pathway. However, the multiple ovulation embryo transfer (MOET) techniques allow greater progress to be made through females because they can give rise to more offspring and thus allow greater selection.

It is the number of animals measured that counts as the population and the fewer that are needed as parents the greater the intensity of selection that is possible. Suppose we have a beef herd in which post-weaning growth is being measured and the average daily gain (ADG) of all bulls is 1.5 kg while that of all heifers is 1.3 kg. Let us assume that a standard deviation is 10% of the mean so that it is 0.15 kg in males and 0.13 kg in females. Let us further assume that we are going to use the best 10% of bulls and the best 30% of heifers to breed cows, but that we will use the best 5% of bulls and the best 10% of heifers as bull mothers. Using Table 5.3 we can calculate how many standard deviations these percentages are above the mean and we already have values for each sex's standard deviation. Table 5.4 shows the selection differentials that would ensue.

If no selection were undertaken in females but the same intensity followed in males then the total selection differential for the four

Table 5.4   Selection differentials by pathway

Route	Percent kept	Merit[a]	Standard deviation (kg)	Selection differential[b] (kg)
Bulls to bulls	5	2.063	0.15	0.309
Heifers to bulls	10	1.755	0.13	0.228
Bulls to heifers	10	1.755	0.15	0.263
Heifers to heifers	30	1.159	0.13	0.151
Total				0.951
Mean selection differential (0.951/4)				0.238

[a] Taken from Table 5.3
[b] Merit × standard deviation

routes would be 0.309 + 0 + 0.263 + 0 = 0.572/4 = 0.143. Failure to undertake female selection would reduce the progress.

These are, of course, predicted selection differentials rather than actuals. They were based on assuming that a given percentage of the population was used and used equally and thus these are the type of calculations one might make beforehand. In reality, even if mated to the same number of females, each male would not give rise to the same number of offspring and this is especially true in litter-producing species. The actual selection differential has to be calculated on the basis of actual use. Table 5.5 shows the results of selecting five bulls with known post-weaning daily gains in a population where the average ADG was 1.5 kg for males. The mean gain of these bulls is 2.0 kg which represents a superiority of +0.5 kg over the population. When they were used and the progeny produced assessed, the true selection differential is calculated by multiplying each bull's superiority by the number of measured progeny produced. This is shown in the final column of Table 5.5 as a value of +0.538. This is slightly higher than the anticipated value assuming bulls were used equally. This is because the better bulls tended to produce slightly more progeny than the poorer bulls which is not always the case.

Table 5.5 Calculation of a selection differential

Bull	Gain (kg)	Progeny measured	Deviation of bull	Weighted deviation
1	2.30	15	+0.80	+12.00
2	2.20	19	+0.70	+13.30
3	2.00	25	+0.50	+12.50
4	1.80	10	+0.30	+ 3.00
5	1.70	11	+0.20	+ 2.20
Total		80	+2.50	+43.00
Mean			+0.50	+ 0.538

The phenotypic standard deviation of a character is the way in which variation is usually described and if the standard deviation is expressed as a percentage of the mean it is called the *coefficient of variation* (CV). Some characters like milk yield are quite variable (they have high CVs) while others like daily gain are less so. Table 5.6

Table 5.6   Phenotypic standard deviations in farm livestock*

Species	Trait	Standard deviation	Units
Dairy cattle	Milk yield	250	kg
	Milkfat yield	40	kg
	Milkfat percentage	0.49	%
	Lactation length	30	days
	ICC milk[a]	125	kg
	ICC fat yield[a]	4.7	kg
	ICC protein yield[a]	3.8	kg
	CGI[a]	96	units
Beef cattle	Birth weight	6	kg
	200-day weight	25	kg
	400-day weight	30	kg
	Pre-weaning gain	0.13	kg/day
	Post-weaning gain		
	feedlot	0.10	kg/day
	pasture	0.08	kg/day
Sheep	Lambs reared	0.6	lambs
	Lamb weaning weight	3.6	kg
	Body weight (hogget)	4.5	kg
	Ewe fleece weight	0.5	kg
Pig	Daily gain	0.06	kg/day
	Feed conversion	0.20	feed/gain
	Backfat (C)	2.5	mm
	Backfat (K)	2.7	mm
	Backfat (S)	4.0	mm
	Carcass weight	1.3	kg
	Dressing percentage	1.60	%
	Litter size	2.8	piglets

* Figures are guides only. Specific populations may vary markedly from these
[a] Holstein-Friesians only
ICC, Improved Contemporary Comparison; CGI, Cow Genetic Index

shows some examples of phenotypic standard deviations for certain traits in farm livestock.

In animal breeding one is usually working with characters which show normal curve features and the superiority of any given percentage of the population can be expressed as the mean standard deviation by which that percentage exceeds the overall mean. This assumes the percentage is taken at one extreme and a brief extract of such values is given in Table 5.3. Extensive tables are given by Pearson (1931) and by Becker (1984) so that Table 5.3 represents only a very small extract. The table assumes a large population because in small populations the values would be different from those shown. Becker (1984) gives extensive tables for small populations of varying sizes.

The importance of fertility in aiding selection is illustrated in Table 5.7, using sheep with different lambing percentages, but assuming equal sex distribution and a 6-year life in the flock. All values given are per 100 ewes mated, but the table could be considered for a flock size of less than or more than 100.

Table 5.7   Female selection intensity in sheep

Lambs weaned per 100 ewes mated	80	100	120	140	160	180	200
Females available	40	50	60	70	80	90	100
Replacements needed[a]	18	18	18	18	18	18	18
Per cent retained[b]	45	36	30	26	23	20	18
Selection intensity (i)[c]	0.88	1.04	1.16	1.25	1.32	1.40	1.46
Relative progress	100	118	132	142	150	159	166

[a] Includes small addition to allow for deaths
[b] Per cent retained of total available
[c] Figures taken from larger version of Table 5.3

As lambing percentage increases more ewe lambs are available for selection but, because the flock size is static, fewer are needed as replacements. Thus, as one goes from left to right in the table the selection that can be applied increases and the superiority of those ewe lambs retained for breeding is enhanced. Relatively speaking, the 200 lamb crop flock (feasible in lowland areas in Britain) is 66% more effective in selection terms than is the case of the flock with an 80% crop (which might occur in poor hill areas in northern Britain). If length of life were to be less than six lamb crops then the advantage of superior lambing percentages becomes even more apparent.

Generation interval

Different species reproduce at different rates. They reach breeding age at different points and have differing gestation lengths. The decision 'when to breed' is partly determined by physiological considerations and partly by man. He can decide to calve a heifer at 2 years of age or wait until 3 years, depending upon economic considerations. This kind of decision has genetic implications because the earlier animals are bred from the quicker one can reach the next generation which, hopefully, will be superior in genetic terms to the present one. In economic terms, the quicker animals are bred from the sooner they start to 'earn their keep'.

Generation interval (GI) is defined as the average age of parents when their offspring are born. In effect it defines the length of time between generations. This will vary between species and populations within species. Puberty sets a lower limit for GI, but the mean value will be higher because animals are bred from on several occasions and are getting older on each such occasion. In man, birth is legally feasible at 16 years of age, but the mean GI is probably closer to 25 years. Values for domestic livestock are shown below (in years):

Horse	9–13	Dog	4–5
Dairy cattle	5–7	Sheep	3–4
Beef cattle	4–6	Cat	3–4
		Pig	1–2

In general, a breeder is interested in progress per unit of time rather than per generation. With domestic animals the time unit is usually on a per year basis. The quicker the generation turnover the more rapid the progress. With traits measured in the live animal, such as daily gain, the generation interval can be kept quite low, but in dairy cattle where bulls do not have producing progeny until they are some years of age the generation interval is raised through waiting for progeny records. MOET schemes can lower this.

Genetic gain

The three components of progress – heritability, selection differential and generation interval – have now been defined. They are combined to predict progress thus:

Genetic gain/generation $=$ heritability $\times$ selection differential

Genetic gain/year $= \dfrac{\text{heritability} \times \text{selection differential}}{\text{generation interval (in years)}}$

These equations are at the core of all improvement. Maximum gain will result when the selection differential (S) and the heritability (h^2) are high and the GI is low. Anything that increases the heritability and/or the selection differential aids progress as does anything that reduces the GI, provided that this does not adversely affect other items among the three.

Measurement of genetic gain

Long-term experiments with domestic livestock are uncommon, especially in terms of generations. In species like mice or *Drosophila*, trials have lasted 30 generations (about 180 years in cattle equivalents!) and the divergence between upward and downward selection has been as much as 20 standard deviations of difference. Similar progress might be feasible in farm livestock given selection for a long enough period of time.

Because genetic improvement can be costly it is desirable to measure progress, if only to assess the returns on expenditure. A simple elevation of the mean may indicate environmental change in the sense of managerial improvements. Dairy cattle average yields in Britain have risen from 3000 to 6000 litres in the past 25 years or so, but not all of this can be attributed to genetics since we have understood more about cattle nutrition and management during this period. There are several ways in which genetic progress might be assessed. One way is to set up control herds/flocks in which selection is undertaken at random and selected lines can then be compared with this control population. If maintained in the same environment or assessed by sire reference techniques or those of best linear unbiased predictions (BLUP) (see Chapter 7) genetic progress can be assessed.

Control populations need to be large enough to allow them to be closed without *inbreeding* levels becoming too high. This means that the number of males needed is higher than would be required in a commercial unit and family sizes are best kept constant. A control herd of pigs once existed at the University of Newcastle upon Tyne based upon 16 boars and 32 sows. Each sow was replaced by a daughter and each boar by a son both selected at random. Eventually the control unit became so far behind the commercial units that comparisons became largely meaningless and the unit itself was always costly, bearing in mind the 'out of date' quality of the pigs. The unit lasted about 14 years before being disbanded.

Another type of control would be to store semen from sires of proven merit and use this at intervals in the population at large. Every 5 to 10 years might be feasible. Such a procedure is easier in cattle than pigs or sheep because of the greater ease in freezing semen. Frozen embryos might also be used to provide similar results.

At present with sire reference schemes and BLUP techniques measurement of genetic gain is becoming easier. An example of this is shown using the Suffolk Sheep Sire Reference Scheme in Britain. This was established using 1990 as a base year. Sheep are assessed on 8-week weight, 140-day weight, muscle depth at 140 days and fat depth at 140 days. These four items are combined into an index where the base value in 1990 was 100. This was assessed for all lambs measured and each year has been reassessed for new lambs born. Table 5.8 shows the index values for each available year since 1990.

Table 5.8   Index values for Suffolk lambs by year of birth

Year of birth	Scheme index
1990[a]	100
1991	106.7
1992	115.3
1993	122.1
1994	128
1995	139.5
1996	146.9

[a] Base year
After Hiam (1997)

It can be seen that in a relatively short time progress has been considerable in index terms and it would be possible to say that the average lamb born in 1996 would be almost 3 kg heavier at 140 days than his counterpart in 1990 and he would be carrying just over 1 mm of extra eye muscle. These values are not apparent from the index which combines four traits but from the component parts not published in this table. Increasingly, index evaluation of this kind is enabling measurement of genetic progress in a variety of species.

In any selection work, limits might be arrived at when the selected criteria reach a plateau after having first slowed down. Usually this means that additive variation has been used up, but if new genes from another population are introduced at this point it may be feasible to advance again. The consequences are shown in Figure 5.1.

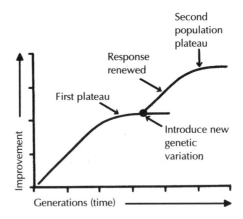

Fig. 5.1   Renewed selection response reaching a plateau at a higher level

With time, additive genes should be being fixed in a selected population and it is generally accepted that heritabilities will slowly decline with time, but the time factor may be quite large. Breeders must also realise that programmes may result in changes that were not necessarily planned. A selection experiment with *ad-libitum* fed pigs at Newcastle and the West of Scotland College was aimed at reducing backfat thickness, improving efficiency and increasing daily gain. The first two objectives were achieved to a marked degree, but daily gain advanced only minimally. It was later found that one consequence of selection was reduced appetite. The pigs ate less, grew at the same rate and were naturally leaner and more efficient.

Chapter 6
Aids to Selection

Pedigree information

A pedigree is simply a record of ancestry and is not synonymous with purebred because a crossbred animal might just as easily have a record of its ancestry. For many years pedigrees, even official ones issued by a breed organisation, were little more than a list of names or numbers and thus had virtually no value in predictive terms. That is still largely true today with cat, dog and, to a degree, horse pedigrees though this may depend upon where one is and what breed one is working with. Dog pedigrees are beginning to include data on coat colours, working traits and some genetic disease tests.

The pedigree in Figure 6.1 is that of a Shorthorn bull called Roan Gauntlet from the early development days of the breed and is typical of the paucity of information at that time and for decades later. This pedigree will be used to measure *inbreeding*, but at present it is shown to illustrate features of pedigrees. Human pedigrees tend to be constructed from the top of the page downwards but animal pedigrees run from left to right. Traditionally sires are listed above dams in any pairing and this would be considered a four-generation pedigree, since the subject's generation is not counted. The tail female line is the bottom line of the pedigree, i.e. Princess Royal – Carmine – Cressida – Clipper and the tail male line would be Royal Duke of Gloucester – Grand Duke of Gloucester – Champion of England – Lancaster Comet. Some breeders still see qualities in this, but there is no special virtue in these two extreme edges of any pedigree. Generations are usually numbered as 1 for parents, 2 grandparents, 3 great grandparents, etc.

Any animal in each generation derives 50% of its genes from each parent and that is valid in each case. Thus, Roan Gauntlet would get 50% of his genes from his sire Royal Duke and Royal Duke would have got 50% of his from each of Grand Duke and Mimulus. On

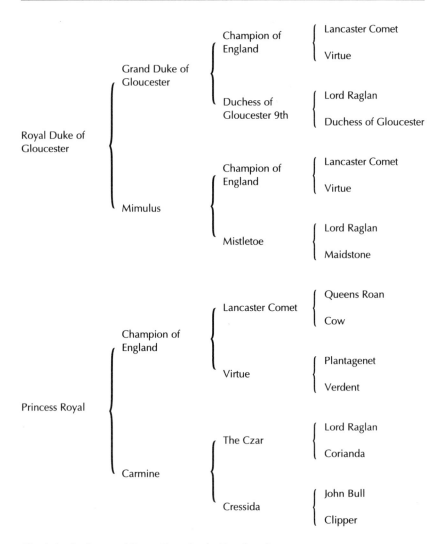

Fig. 6.1 Pedigree of Roan Gauntlet (a Shorthorn)

average, therefore, Roan Gauntlet should have 25% of his genes from each of Grand Duke and Mimulus, but that is not necessarily the case though it is a good guide. It does not follow that the 50% Roan Gauntlet obtained from Royal Duke contained equal proportions from Royal Duke's two parents, still less ancestors further back.

Clearly, the nature of pedigrees is such that the number of ancestors doubles each generation back. By generation six there are 128 ancestors, but a bull only has 60 chromosomes. Some ancestors may have contributed only a part of a chromosome to their sixth generation descendant or they may exist in the pedigree without any of their genes having ended up in their descendant. The further back one goes the less importance any ancestor can have and the less bearing they have upon the animal being studied.

Nomenclature varies with species. Most breeders in any species have an affix which usually becomes the first name of all animals bred by that breeder, e.g. Hunday, Terling, Vikkas. A dog breeder would probably name all animals in a litter with the same letter, e.g. Vikkas Alaric, Vikkas Arno and gradually work through the alphabet. In farm livestock, especially cattle, attention is given to what are called 'female families'. The term family is rather misleading. For example, a family might begin with a cow called Pearl and all her female descendants will be called Pearl with a number added. Thus, Pearl 144th would be the 144th Pearl born and would trace back in tail female line to the original Pearl. However, Pearl 144th might be a paternal half-sib to Buttercup 106th and thus have more in common with that cow than any other Pearl in the herd. Some herds incorporate the sire's name into the cow's name. The MOET herd in Northumberland uses this technique, so that MOET Besne Cadet E was sired by Besne Buck out of MOET Winken Cadet K. In this herd there might be a number of MOET Besne Cadets but they would each have a different letter appended.

Families of the Pearl-kind cited may not be very meaningful in genetic terms and family selection in a genetic sense should not be confused with this naming system. Similarly, the fact that animals carry a common affix or herd name does not make them a 'line' or 'strain' or even indicate any relationship. Nomenclature such as the MOET herd use does help in remembering parentage but may have little genetic significance.

Official breed society pedigrees exist as a record of ancestry, but evidence suggests that between 5 and 10% of pedigrees may be inaccurate through mistakes, however innocently made, about ancestry. Most AI stud organisations now do DNA checks on stock to ensure that ancestry is what it claims to be. However, that may not always be feasible in commercial units, especially those using natural service or extensively ranching.

The value of a pedigree depends upon the information it contains and will depend upon the closeness of the relationship of the animal and its ancestors. It also will depend upon the heritability since the higher this is the greater the value of pedigree data. By the same token, the higher the heritability, the more useful information will be on the subject's own performance.

Table 6.1 shows the accuracy of selection of pedigree records and of records on the subject itself. In this table and in similar tables (6.5 and 6.6) it is assumed that data are based upon a single record for each animal involved, as opposed to multiple records. In Table 6.1 increased heritability increases the value of all types of record, but within each, heritability information on a full three-generation pedigree is more valuable than just parents and grandparents and that, in turn, is more useful than just parents. This is to be expected, but it must also be noted that data on the individual itself are always more valuable than pedigree data and the higher the heritability the more valuable the individual's data relative to pedigree. Modern pedigrees are becoming more detailed in terms of information. They also tend to be published in a curtailed form with just one or two generations. A dairy bull pedigree would contain lactation yields of the dam and granddams plus fat and protein percentages or yields as well as some measures of merit, e.g. Profit Index (PIN) numbers (see Chapter 9) for all immediate ancestors.

Of course, one has a pedigree before one has an individual born with that pedigree, in the sense that a breeder can (and very often may) construct 'potential' pedigrees on the basis of mating sire A to either dam B or dam C. It is possible to predict what is called an

Table 6.1 Accuracy of selection of pedigree records (%)

Heritability (%)	Pedigree records on the			Subject's own performance
	Parents only	Parents plus grandparents	Complete pedigree	
10	22	27	29	32
20	32	37	39	45
30	39	43	45	55
40	45	49	50	63
50	50	53	54	71
60	55	57	57	77
70	59	61	61	84
80	63	64	64	89

estimated breeding value (EBV) for an individual on the basis of pedigree information. If data exist on the animal itself then its EBV can be calculated from the heritability of the trait multiplied by the deviation of that individual's performance from that of contemporaries. Pedigree prediction is more reliable the greater the information, i.e. as one moves down Table 6.1 and from left to right.

Performance testing

Performance testing involves recording the performance of the animal(s) under study and making decisions in the light of those data. A geneticist might call this individual or mass selection, but animal breeders refer to it as performance testing.

The usefulness of performance data increases as the heritability increases (see Table 6.1). In fact, the accuracy of a single record on an individual is the square root of the heritability. In practical terms animals being considered for selection are usually tested on some particular management regime with all animals undergoing the test more or less simultaneously, i.e. they are contemporaries. If beef bulls were being tested on post-weaning growth measured from about 200 to 400 days of age then these bulls would be weaned as close to 200 days as possible and placed in a particular unit and fed the same diets. Complete diets would ensure that all bulls ate the same composition of diet and ideally group housing with individual feeding facilities (such as Callan-Broadbent feeders) would be preferable to individual housing. Even if this is done meticulously, pre-test environmental effects, especially with weaning as late as 200 days, could influence performance on test. If bulls came from heifer dams they might be less well grown and might compensate on test. Similarly, bulls reared exceptionally well prior to going on test might perform less well on that test. Corrections for dam age/parity as well as starting age may be required.

In general terms, performance tests are intended for individuals tested at the same time and place, on the same regime. The term 'same time' may require some latitude. Testing at special centres is easier than on-farm testing but more costly and is gradually falling into disfavour. At the same time, on-farm testing, which used to suffer from a paucity of numbers of animals in many instances, has been given new impetus by the use of *reference sire* techniques and BLUP methods.

In general terms, performance testing is only really suitable for traits of moderate to high heritability and is of no use for traits which cannot be measured in that sex. Dairy bulls are not performance tested, other than for growth or conformational traits, but beef bulls, sheep and pigs can be frequently tested on their own performance. In some instances carcass traits can also be assessed without having to slaughter the animal, and while, at one time, this applied principally to pigs it is now increasingly used in cattle and sheep. Measuring the ability to do something, e.g. racing performance in racehorses or greyhounds, or herding ability in sheepdogs, is also an example of performance testing. It must be realised that the measurement itself can add variability to the character being studied. The more objective and uniform the way measurement is made, the more valuable will be that measurement. Subjective measurements can, nevertheless, be valuable in performance testing.

Multiple records

Performance testing usually relates to one record, as a bull, for example, can only be assessed once on its growth from 200 to 400 days. However, some animals can be assessed more than once for specific traits. We can measure greasy fleece weight each time a sheep is shorn and lactation yield or litter size can be measured more or less annually. This means that we can measure the degree to which performance is repeated. This allows an estimate of the variation between animals and also within individual animals. The correlation between different or repeated measurements of the same animal is termed the *repeatability* and can range from 0 to 100%.

Repeatability measures the extent to which an animal's superiority (or inferiority) is repeated in successive years or lactations. If a character were highly repeatable an animal would keep showing the same performance. The extent to which performance is repeated will depend upon the permanent differences between individuals. Permanent differences include those that are genetic and those caused by permanent environmental features. The repeatability is defined as:

$$r = \frac{\text{genetic variance} + \text{general environmental variance}}{\text{total phenotypic variance}}$$

which in terms of symbols becomes:

$$r = \frac{\sigma_A^2 + \sigma_D^2 + \sigma_I^2 + \sigma_{E_g}^2}{\sigma_A^2 + \sigma_D^2 + \sigma_I^2 + \sigma_{E_g}^2 + \sigma_{E_s}^2}$$

The numerator and denominator in this equation are identical except for the specific environmental variance. If the repeated records of a character are going to fluctuate markedly because of features peculiar to each new measurement then specific environmental variance is likely to be high and repeatability will be on the low side. If such special environmental variance is minimal then repeatability will tend to be high and denominator and numerator will become more similar.

The heritability measured the additive variance as a proportion of the total phenotypic variance. Thus, there is a similarity between h^2 and r, but the latter measurement will be higher in each case unless dominant, epistatic and permanent environmental variances were all zero. Since this is improbable we can say that the repeatability sets a kind of upper limit for the heritability.

By taking repeated measurements the breeder is reducing the importance of the specific environmental variance. Two measurements would mean halving this value. Continual recording of performance, for example each lactation, gives rise to what is called *lifetime performance* and the average of these records would be average lifetime performance. Using multiple records can reduce total phenotypic variance by reducing special environmental variance and may increase the additive variance as a proportion of the total. A measure of this is given by the formula:

$$\frac{n}{1 + (n - 1)\,r}$$

where n is the number of records and r is the repeatability. Table 6.2 shows this formula worked out for different numbers of records and repeatabilities.

Let us look at an example of using Table 6.2. Suppose on one record a character has a heritability of 0.3 and a repeatability of 0.5. If

Table 6.2 Equation $\dfrac{n}{1 + (n-1)\,r}$ worked out for various numbers of records and repeatabilities

Number of records	Repeatability								
	0.1	0.2	0.3	0.4	0.5	0.6	0.7	0.8	0.9
1	1.00	1.00	1.00	1.00	1.00	1.00	1.00	1.00	1.00
2	1.82	1.67	1.54	1.43	1.33	1.25	1.18	1.11	1.05
3	2.50	2.14	1.88	1.67	1.50	1.36	1.25	1.15	1.07
4	3.08	2.50	2.11	1.82	1.60	1.43	1.29	1.18	1.08
5	3.57	2.78	2.27	1.92	1.67	1.47	1.32	1.19	1.09
6	4.00	3.00	2.40	2.00	1.71	1.50	1.33	1.20	1.09

two records are used the heritability would have to be multiplied by 1.33 (see Table 6.2) and thus become 0.40. With three records it would become 0.45 and with six 0.51.

The values in the body of the table increase from top to bottom, i.e. as more records are used, but decline from left to right as repeatability increases. In effect, Table 6.2 illustrates the advantage (or not) of waiting for more records. If a repeatability is low, and therefore the heritability is also low, more records would increase the heritability by relatively large amounts though it would still remain relatively low. In contrast, waiting for more records when the repeatability is high is less useful because a high repeatability means that one record is a good guide to performance and waiting for more records would be merely time-consuming for minimal gain.

Lifetime records are useful in identifying those animals which have proved their ability to both survive and perform well. However, the need to wait for more data increases not only the reliability of records and the heritability but also the generation interval which could, overall, produce a slower rate of genetic progress per unit of time.

Table 6.3 shows a simple breeding value assessment for four cows with varying numbers of records based on weaning weight of their calves. It is assumed that heritability is 0.3 and repeatability 0.5. The deviation of each calf's weaning weight from the herd average is given and the total and mean deviations are then calculated. Reliability indicates the value of the heritability based upon numbers of records, using Table 6.2. Cow A stays at 0.3 because it has only one record, but cow B becomes 0.45 (0.3 × 1.5) and cows C and D become 0.48 (0.3

Table 6.3 Breeding value assessments for weaning weight of beef cattle (hypothetical)

Calving	Calf weight as deviation from herd mean (kg)			
	Cow A	Cow B	Cow C	Cow D
1	−20	+ 50	−10	+ 35
2		+ 30	+15	+ 30
3		+ 40	+25	+ 35
4			+20	+ 40
Total deviation	−20	+120	+50	+140
Mean deviation	−20	+ 40	+12.5	+ 35
Reliability[a]	0.30	0.45	0.48	0.48
Breeding value	− 6	+ 18	+ 6	+ 16.8

[a] Reliability is derived from Table 6.2 assuming a heritability of 0.3 and a repeatability of 0.5

× 1.6). The breeding value then becomes the mean deviation multiplied by these reliability figures.

The order of merit would be cows B, D, C and A and certainly B and D seem to be above average whereas C is not particularly outstanding. Cow A might have had some problem with her sole calf and this may not be repeated with a second calf. Simple breeding values of this kind are predictions, not exact features, but they are more accurate the more information they are based upon.

Some measures of repeatability are shown in Table 6.4.

Progeny testing

Decisions can be based upon the performance of progeny rather than upon that of the subject. This is usually applied to males because females tend to have fewer progeny, but in litter-species, testing of females may be feasible and with increasing use of MOET techniques more can be done with females. Some traits are, of course, not measurable in males and thus progeny testing is feasible while performance testing is not.

Progeny represent a sample of the genes of their father (and mother) and, given enough samples, will give a reasonable idea of the superiority (or otherwise) of that sire. This is, of course, only possible

Table 6.4   Repeatability of farm livestock traits (%)

Species	Trait	Repeatability
Dairy cattle	Services per conception	12
	Annual non-return rate	6
	Bull ejaculate volume	70–80
	Milk yield	40–60
	Milkfat percentage	40–70
	Milking rate	80
Beef cattle	Birth weight	20–30
	Weaning weight	30–55
	Yearling weight	25
	Daily gain to weaning	7–10
	Body measurements	70–90
Sheep	Ovulation rate	60–80
	Lambs born/ewe mated	15
	Lambs born/ewe lambing	30–40
	Lambs weaned/ewe mated	18–20
	Birth weight	30–40
	Lamb gain	38–50
	Fleece weight	30–40
	Various wool traits	50–80
Pigs	Litter size/birth	10–20
	Litter size/weaning	10
	Litter weight/birth	25–40
	Litter weight/8 weeks	5–15
	Birth weight/pig	20–40
	Weaning weight/pig	10–15
	Adult live weight	35
	Eye muscle area	95

if the progeny of a particular sire are compared with the progeny of other sires in the same environmental conditions. The performance of a bull tells us what he *might do* if used whereas the progeny test tells us what he *is doing*.

Progeny testing is time-consuming and increases the generation interval. For example, a dairy bull will be about 6 years of age before he has sufficient progeny lactating to allow estimates of his breeding worth for milking traits. Performance testing is thus preferable whenever it can be performed, which, in this example, is not possible.

Usually, progeny testing is undertaken for traits of low heritability and for which performance is thus a poor guide. It is, obviously, used for traits expressed in one sex (e.g. milk production) and for those measurable only after death (e.g. some carcass traits).

To be effective, progeny tests require:

(1) As many sires on test as possible (five to ten minimum).
(2) The random mating of dams to ensure that specific sires are not mated simply to very good or very poor dams and that sires get similar numbers of young/old mates.
(3) As many progeny per sire to be produced as possible (ten minimum for some traits but 200 to 300 for others such as dystocia).
(4) Progeny to be unselected during the test.
(5) Progeny to be treated in the same way or comparisons made within the herd/flock and year/season groupings.

Given these standards, progeny testing can be very effective and the most accurate way of measuring breeding merit. Obviously the reliability increases with heritability and the number of progeny being tested so that the figures in the body of Table 6.5 increase from top to bottom and from left to right. However, given high numbers of progeny little extra accuracy is gained by increasing numbers. Progeny testing is considerably better in accuracy terms than pedigree

Table 6.5 Accuracy of selection of progeny records (%)

Heritability (%)	Number of progeny tested							
	5	10	20	40	60	80	100	120
10	34	45	58	71	78	82	85	87
20	46	59	72	82	87	90	92	93
30	56	70	79	87	91	93	94	95
40	60	73	83	90	93	95	96	96
50	65	77	86	92	95	96	97	97
60	68	80	88	94	96	97	98	98
70	72	82	90	95	96	97	98	98
80	75	84	91	95	97	98	98	98

data and with low heritabilities as few as five random progeny can be as accurate as performance data. However, at low heritabilities progeny numbers become crucial.

Practical comparisons

Let us assume that we are looking at dairy bulls and that we have available a total of 2000 milk-recorded cows on which testing can be undertaken. Assume a heritability of milk yield of 0.30 and that ten progeny-tested sires are required each year. Obviously one can test larger numbers of bulls on smaller numbers of progeny or vice versa.

Table 6.6 shows the number of bulls that could be tested assuming different progeny numbers per bull. On the left of the table a small number of bulls is tested on large progeny numbers and as one moves to the right the number of bulls increases and the progeny per bull declines. The percentage of bulls retained is based upon keeping ten sires in each situation. The superiority (S) is taken from an extended version of Table 5.6 and the accuracy or reliability (R) is taken from Table 6.5 based on progeny numbers. The product of S × R shows the relative value of the different methods. In terms of effectiveness based on this heritability it is better to test about 100 bulls on 20 progeny each than any other combination. In other words it is better to test large numbers of bulls somewhat less efficiently and retain a small proportion than test very accurately and retain a high proportion.

Table 6.6   Number of bulls versus reliability of the test

Bulls tested	16	20	25	40	80	100	166
Progeny/bull	125	100	80	50	25	20	12
Per cent kept	63	50	40	25	13	10	6
Superiority (S)	0.60	0.80	0.97	1.27	1.63	1.76	1.99
Reliability (R)	0.95	0.94	0.93	0.90	0.82	0.79	0.70
S × R	0.57	0.75	0.90	1.14	1.34	1.39	1.39

Information from other relatives

Sometimes it is possible to obtain data from relatives such as full-siblings (brothers and sisters) or half-siblings (usually by the same sire). The accuracy of selection using these records is shown in Table 6.7 which can be compared with Tables 6.1 and 6.5.

Table 6.7 Accuracy of selection of sibling information (%)

Heritability (%)	Number of full-sibs			Number of half-sibs			
	2	4	8	5	10	20	40
10	22	29	38	17	23	29	36
20	30	39	48	23	29	36	41
30	36	45	54	27	33	39	44
40	41	50	58	30	36	41	45
50	45	53	60	32	38	43	46
60	48	56	62	34	40	44	47
70	51	58	63	36	41	45	48
80	53	60	65	37	42	46	48

In broad terms the same principles apply as in the companion tables in that as heritabilities and sibling numbers increase accuracy also increases, but full-sibs are more useful than half-sibs and full-sibs are more accurate for testing than pedigree data (see Table 6.1). The use of sibling data has come into prominence since the development of MOET techniques (see Nicholas and Smith, 1983). This will be discussed in greater detail later (see Chapter 9), but it is useful to point out here that the MOET system has been most successful in dairy cattle. MOET depends upon the production of multiple embryos and their implantation into donor dams. Sires can then be evaluated on their full- and half-sister performance at a much younger age than is feasible with conventional progeny testing. Sibling testing is not as accurate as progeny testing but it can be done much faster, resulting in progress on a per year basis that could equal or exceed progeny testing.

Female families were mentioned earlier in this chapter (see 'Pedigree information') and this nomenclature system in cattle and pigs can be confusing when talking about family selection. Falconer and Mackay (1996) have looked at the selection on an individual, family or within-family basis. In this context families are intended to be either collections of full-sibs or alternatively half sibs.

Figure 6.2 represents four situations A, B, C and D, with the numbers 1 to 5 signifying five families (full- or half-sibs) in each instance. Merit is based on phenotypic merit and is greatest near the top of the figure. In each population represented by the letters there

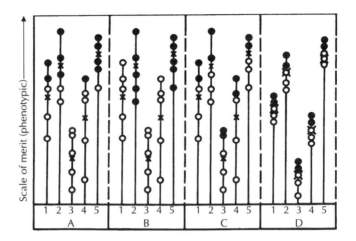

Fig. 6.2   Different methods of family selection (redrawn from Falconer and Mackay, 1996)

are 25 individuals. With individual selection (A), merit, assuming ten individuals are required, would be based upon the best ten animals regardless of location so the ten nearest the top of the figure are selected. With family selection (B), the ten individuals chosen would be based on the overall merit of the families (family mean) and involve taking all individuals of the two best families (2 and 5). Using within-family selection (C), the ten individuals chosen would comprise the best two in *each* of the five families. The chosen individuals vary widely between A, B and C, but in general terms A would be the wisest choice and individual selection would turn out to be the best method. One problem with family selection is that whole families are chosen and discarded. This reduces the lines within the population and could lead to inbreeding. Choice D represents a situation in which there are large differences between families but small differences within families and in this situation within-family selection can be useful. See Falconer and Mackay (1996) for more mathematical discussion on this issue.

Combined selection

It is possible to collect information from several sources, pedigree, siblings, the individual itself and eventually progeny of that individual. Such data can be combined into an index which produces a single figure result, e.g. 100 = average. Clearly, information has to be weighted according to its value and reliability or accuracy. The index may look like this:

$$I = b_1 A_1 + b_2 A_2 + b_3 A_3 + b_4 A_4 + \ldots$$

where b represents a constant, A represents the phenotypic value of trait A and subscripts 1 through 4 represent the various items of information: family, individual, etc. Such indices may be difficult to calculate but are relatively easy to use since the breeder uses the highest index values (I).

Initially such an index might be based upon sibling or pedigree data and the performance of the individual, but gradually progeny data can be added in (see Nicholas, 1987; Falconer and Mackay, 1996; Cameron, 1997).

Chapter 7
Selection for Multiple Objectives

Introduction

Thus far, breeding and selection have been described in terms of single characters. In practice, few breeders select for only one thing. Livestock breeders usually have a series of objectives which will include specific production traits as well as, perhaps, some aspects of physical appearance. The more things one seeks to obtain, the harder it will be to achieve one's objective and breeding is best directed towards objectives that are feasible in any given situation. Selection for multiple objectives does require that there is information available on certain genetic parameters if success is to be achieved.

Correlation between characters

Breeders are well aware that certain characters seem to be related to others. Selection for increased wither height would be likely to lead to increased body weight. Selection for milk yield is likely to lead to a reduction in milkfat content of that yield. Breeding for increased feed efficiency in meat animals will, probably, be associated with leaner carcasses.

Beliefs about relationships between traits may not always be borne out. Not all blondes have blue eyes! Nevertheless, there are relationships between some hair and eye colours so that they appear together more frequently than might be expected at random. Pronounced milk veins on an udder do not necessarily imply high milk yields any more than thin skin denotes high milk potential. Some popular beliefs about relationships are without foundation while others have some basis in fact and a biological explanation.

Valid relationships stem from physiological ones and it is known that genes which affect a particular character may also affect other characters, just as environments which affect one character in a specific way may affect others. Relationships between traits are called correlations and can range from -1.0 to $+1.0$, with the former indicating a high negative relationship and the latter a high positive one. Negative relationships mean that as one trait increases the other decreases while with positive relationships both traits move in the same direction. Negative relationships are not necessarily undesirable. A negative relationship between backfat thickness and feed efficiency in pigs would be desirable because as efficiency increased (improved) backfat would reduce (improve).

If two characters are measured in a series of individuals then the correlation between them is called the phenotypic correlation (r_P). Examples of such correlations are shown in Figure 7.1. This *phenotypic correlation* can be divided into genetic and environmental parts which are termed r_G and r_E respectively. *Genetic correlations* exist because the same genes affect both characters (a phenomenon called *pleiotropy*) while *environmental correlations* exist because specific environmental conditions produce certain features in animals which result in a correlation between these features and the environmental conditions.

In animal breeding terms one is usually interested in the relationship between breeding values so the genetic correlation is best thought of as r_A while r_E is the environmental correlation including non-additive genetic effects, in other words everything except the additive factors. The calculation of genetic correlations involves collecting data on groups of relatives such as half-sibs or parents and offspring. Usually there are large sampling errors and as a consequence genetic correlations are not always very precise. Moreover, since they depend upon gene frequencies genetic correlations of the same traits in different breeds or populations of the same breed could differ markedly. Phenotypic correlations are easily calculated, but all correlations benefit from being assessed on large numbers.

Tables 7.1 to 7.4 show examples of correlations and in these the genetic correlations appear above the diagonal line and phenotypic ones below the line. It should be realised that the data in these tables are illustrative guides. Alternative results will ensue from different populations so too much should not be read into these values.

Genetic correlations are useful to a breeder. If two characters are related, the consequence of increasing one upon response in the other

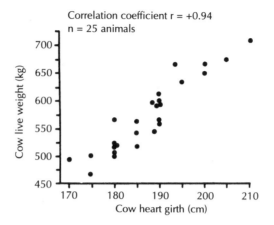

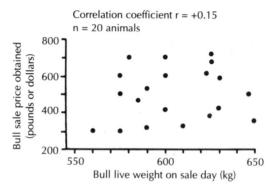

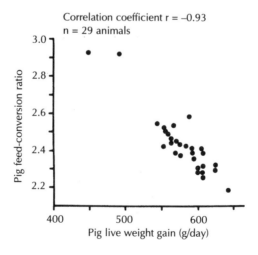

Fig. 7.1 Examples of phenotypic correlations between traits

Table 7.1 Correlations among traits in dairy cattle

Trait	MY	BY	PY	FP	PP
Milk yield (MY)		0.82	0.87	–0.27	–0.18
Butterfat yield (BY)	0.88		0.86	0.26	–0.11
Protein yield (PY)	0.95	0.93		0.04	0.22
Fat % (FP)	–0.20	0.24	–0.01		0.55
Protein % (PP)	–0.19	–0.04	0.06	0.49	

Figures above the diagonal are genetic and those below are phenotypic correlations
Source: Gibson (1987) who drew from various sources

Table 7.2 Correlations among traits in beef cattle

Trait	BW	WW	YW	MS	FD
Birth weight (BW)		0.63	0.56	0.38	N/A
Weaning weight (WW)	0.65		0.76	–0.56	0.19
Yearling weight (YW)	0.39	0.74		–0.65	0.10
Muscling score (MS)	0.08	–0.07	–0.01		0.08
Fat depth (FD)	N/A	0.19	0.29	–0.09	

Figures above the diagonal are genetic and those below are phenotypic correlations
Source: Mohiuddin (1993)
N/A, not available

Table 7.3 Correlations among traits in Merino sheep

Trait	BW	GFW	FD	SL	C
Body weight (BW)		0.26	0.12	0.04	0.40
Greasy fleece weight (GFW)	0.36		0.19	0.70	–0.20
Fibre diameter (FD)	0.13	0.13		0.44	–0.10
Staple length (SL)	0.10	0.30	0.11		–0.34
Crimps/inch (C)	0.05	–0.21	–0.13	–0.22	

Figures above the diagonal are genetic and those below are phenotypic correlations
Source: highest values of Newton Turner and Young (1969) who drew from various sources

can be assessed. For example, selecting bulls on 400-day weights
would be quite successful because the character is quite highly
inherited. However, an increase in 400-day weight would involve
some increase in weight at other ages because of positive genetic
correlations. Birth weight might well increase and bring with it
greater dystocia. Knowing this the breeder could select heavier 400-
day weights but use bulls which were known to give fewer problems at

Table 7.4 Correlations among traits in pigs

Trait	DG	FE	KO	CL	BF	EMA
Daily gain (DG)		−0.76	−0.19	0.14	−0.15	−0.11
Feed efficiency (FE)	−0.73		0.01	−0.08	0.21	−0.34
Killing out % (KO)	−0.17	−0.05		−0.40	0.28	0.36
Carcass length (CL)	0.07	−0.04	−0.19		−0.30	−0.08
Backfat thickness (BF)	−0.07	0.19	0.19	−0.22		−0.28
Eye muscle area (EMA)	−0.03	−0.16	0.15	−0.05	−0.13	

Figures above the diagonal are genetic and those below are phenotypic correlations
Source: Dalton (1985)

calving. This latter aspect would require large-scale progeny testing to evaluate.

Then again, knowing that characters are related, there may be an opportunity to select for one character and bring about a related or correlated response in a second character which might be much harder or more costly to measure directly. Usually, selection is more successful if made directly for the character in which one is interested. However, there can be occasions when selection for a correlated character can bring about greater progress in the trait concerned than can direct selection. This is particularly true for characters that cannot be measured in both sexes or which are very costly to assess. Suppose we examine the response to direct selection for trait A, which we can designate R_A, and then look at the correlated response in A, which we can designate CR_A, by selection for trait B. The consequences will depend upon the respective heritabilities of A and B, the intensity of selection possible in A and B and the genetic correlation between them. This is expressed as:

$$\frac{CR_A}{R_A} = R_A \frac{i_B}{i_A} \sqrt{\frac{h_B^2}{h_A^2}}$$

Suppose selection was for litter size ($h^2 = 0.10$) or for yearling weight ($h^2 = 0.35$) in sheep with a genetic correlation between them of 0.17. Let us further suppose that 10% of males and 40% of females are to be selected. We can assess the direct results of selection for litter size and the indirect results upon litter size of selecting for yearling weight. Note that with litter size the trait cannot be measured in males so that selection by this pathway is nil. Table 7.5 shows the selection differentials as taken from a larger version of Table 5.3.

Table 7.5 Selecting for litter size by increasing yearling weight in sheep (hypothetical)

	Selection differentials for	
	Litter size	*Yearling weight*
Females	0.966	0.966
Males	0	1.755
Average	0.483	1.360

The relationship between direct and indirect selection is shown below:

$$\frac{CR_A}{R_A} = 0.17 \times \frac{1.360}{0.483} \times \sqrt{\frac{0.35}{0.10}} = 0.90$$

In this case, correlated response would be 90% of that achieved by direct selection. This is quite a high value and results from the greater selection possible in yearling weight (measurable in both sexes) than in litter size. Given a greater genetic correlation, even higher values would result. In this instance, correlated response (CR) did not outperform direct selection (R), but in some instances the CR:R ratio can exceed unity.

Tandem selection

The simplest way of selecting for many features is to select each in turn. Thus, A is selected for some generations and then B is selected for, then C and so forth. If the characters involved are uncorrelated then selection for A will leave B and C unchanged. If the traits are positively correlated then selection for A will lead to some response in B and C. However, if the traits are negatively related, then as A improves, B and C become worse. The time spent on each trait would depend upon its importance, but the procedure is lengthy and the least useful of the techniques available and hence rarely used in practice.

Independent culling levels

The technique of independent culling involves setting a minimum level of performance for each trait of interest. Any animal failing to meet this minimum in *any* trait is automatically culled. The system has the advantage that it can be operated even with a large number of traits and selection is simultaneous for all traits. However, it has the drawback that an animal which excels in a specific trait but which fails, however slightly, in another would be culled. The procedure using two traits is shown in Figure 7.2.

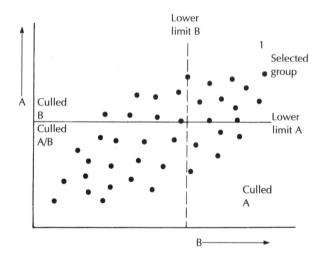

Fig. 7.2   Independent culling levels illustrated for two traits

Individuals in segment 1 are retained and all others culled. As with any of the techniques there is a need to limit the number of traits being selected and to set realistic targets. Nothing is gained by setting targets that few animals achieve, especially if these are for minor features and in reality the targets are going to be set in the light of how the population performs and how many breeding stock are needed. The technique operates better than tandem selection, but with modern computer techniques it is little used in farm livestock. It is probable that many dog and cat breeders use this method, even if they do not call it by this name. They are selecting for a number of features and picking those animals which achieve specific standards in them, usually in respect of conformation and mental attributes. Unfortunately in these species the heritabilities of most traits are unknown or have been evaluated in a small number of breeds for a few characteristics.

Selection index

The *selection index* technique involves pooling information on the individual traits being considered into a single value called an index. It is a form of multiple regression analysis based upon breeding values for the various traits. The index takes the form of:

$$I = b_1X_1 + b_2X_2 + b_3X_3 + b_4X_4 + \dots$$

where the b values are regression coefficients and the Xs are the phenotypic values of the traits measured as deviations from the overall mean. The use of an index is relatively simple. Often they are calculated from a base year, with a value of 100 being the base, and thus the higher the value the better the potential of the animal bearing it. Calculation of indices is, however, quite complicated. Certain data are needed:

- heritability of each trait being considered;
- phenotypic variance of each trait;
- genetic and phenotypic correlations between traits;
- relative economic value of each trait.

The aim of the index is to give the best prediction of the animal's overall merit by pooling information in the best possible way. The animal which excels in one trait but is slightly deficient in another would be culled under the independent culling levels system but not necessarily in an index technique. Correctly made, the index maximises the correlation between true breeding value and the prediction of breeding value or accuracy of the evaluation.

At the start of any selection programme not all the data needed may be available. Genetic parameters may not be known for the population under study and established (and possibly unreliable) figures from other populations may have to be used until the programme itself generates data. Economic value may present some difficulties since it may not be a constant feature but alter as the costs of inputs and the value of outputs change and this could necessitate modification of an index. Like all systems, indices can only be applied to the population being evaluated. In broad terms indices are better than or equal to the independent culling technique and the more traits that are included the more likely is the index to be superior. Further explanations are to be found in Falconer and Mackay (1996).

Specialised lines

A fourth way of selecting for multiple objectives is to develop specialised lines. Usually such lines are usually intended for meat purposes and the objective is that the slaughter generation is a cross of these lines. Often these are male or female lines. Male lines would be selected for features such as growth, feed efficiency and carcass traits while female lines would be selected for reproductive traits and maternal aspects. Crossing the two exploits *heterosis* (see Chapter 8, 'Crossbreeding').

To a degree, specific breeds could be termed specialised lines. In beef cattle, for example, Charolais or Simmental breeds are usually used as terminal sire breeds rather than as the female side of a cross, where smaller crossbred cattle may be preferred. Similarly, in sheep, the Suffolk in Britain is a terminal sire breed used on such F1 crosses as Mules. In pigs in Britain dam lines usually comprise some combination of Large White and Landrace while male lines may contain Hampshire or Duroc material in addition to these two white breeds.

Best linear unbiased predictions (BLUP)

Best linear unbiased predictions (BLUP) were first developed by Henderson (1949, 1973) and are probably the most powerful selection tool. Although the principles were laid down many years ago they have only come into operation in the past two decades through the development of more powerful computing systems. They can be used to provide breeding value estimates on animals born in different locations and years and can take account of environmental trends over time as well as relationships between sires being evaluated. With numerous herds, years and seasons it becomes a major statistical exercise to seek to assess the relative importance, still less the merit, of the individuals being measured. The use of common sires (termed reference sires) in component herds or flocks allows BLUP techniques to be used to assess herd, year and season effects as well as calculate genetic trends, biases due to culling or selection and evaluate sires used in different locations. It is beyond the scope of an introductory book such as this to look at BLUP techniques in detail, for which readers are referred to Nicholas (1987), Cameron (1997) and other books in the Further Reading section. BLUP techniques are now being used in the evaluation of dairy bulls, sheep and other species as the best technique.

Selecting for threshold traits

When threshold traits are being considered, selection is made difficult because phenotypic appearance gives no clues to the underlying genetics, except with individuals which show the threshold trait. If a threshold trait appears in, say, 10% of the population, and one is selecting against it then taking 15% of the normal (unaffected) population would not be equivalent to breeding from the 'best' 15% because one would be taking merely 15% (at random) from the 90% normals.

Selection in such cases is difficult and family selection may be more advantageous than individual selection. Taking into account family/ pedigree data may allow selection of those individuals which not only are normal but also do not appear to be closely related to affected animals or stem from families in which the incidence of the threshold trait is much lower than average. This may be a rather empirical method of selection but there are no real alternatives. Threshold defects which are important need to be recorded in order that family incidence can be estimated and selection against the defect more accurately undertaken. Selecting animals unrelated or less closely related to affected stock or from families with a low incidence will result in a gradual decline of the threshold trait assuming its heritability is moderate to high.

Genotype–environment interaction

One problem met with in animal breeding is that of where and in what conditions to test animals. For example, will performance testing of beef bulls on high concentrate diets be useful in identifying those bulls whose progeny will do well on pasture or on a silage system? Will pigs doing best on *ad-libitum* diets be the pigs which will do best on semi-restricted diets? Will pigs selected in housed conditions be suitable for use in outdoor situations?

All of these questions and many similar ones are concerned with genotype–environment interactions. Confusion arises if breeders believe that effects upon the animal's phenotype also influence the genotype. Cattle fed on pasture are unlikely to grow as quickly as those fed on high cereal diets, but this is a phenotypic effect. The genotypes of the cattle are unchanged in the same way that artificially removing horns will not lead to naturally polled progeny in the next generation.

A genotype–environment interaction exists when the ranking order of individuals changes in different environments. If on pasture, the ranking order of seven bulls based on progeny (or alternatively seven different breeds) was 1, 2, 4, 5, 6, 3, 7, but on concentrate diets the ranking order was 7, 3, 6, 5, 2, 4, 1 and there would be an almost complete reversal of merit order. It would be unwise to select other than in the environment in which the bulls (breeds) were to be used.

At one time it was suggested that animals should be selected in the best possible environment in order that they could express their potential to the fullest extent, but the existence of genotype–environment interactions quickly negated this theory. Experimental evidence suggests that dairy and beef cattle have few important genotype–environment interactions but that they appear to be more important in pigs and sheep. One cannot predict where interactions will occur, but their possible existence needs to be borne in mind.

If there is an important interaction then animals may be best selected in the environment in which they will be used. When stock are used in widely differing environments, climates or management regimes then breeders have to be aware of their possibilities. The old adage 'horses for courses' has genotypic–environmental connotations. Few breeders would use an early maturing beef breed on a high energy diet system since the animals would reach a sufficiently high carcass fat level at too low a body weight. One would be better using later maturing breeds on such a system since they would reach higher weights before reaching a suitable fat level.

Chapter 8
Breeding Systems

Introduction

Having decided which animals to breed from, breeders must decide how these will be mated. In broad terms breeders can decide whether to mate animals of the same breed or different breeds and, if choosing the first policy, can decide whether to mate related or unrelated individuals. The possibilities are summarised in Table 8.1.

Table 8.1 Breeding systems that can be considered

Mating different species	Mating different breeds	Mating the same breed
Species crossing	Crossbreeding	Mating relatives
	Grading-up	inbreeding
	Backcrossing	linebreeding
	Crisscrossing	Mating unrelated stock
	Rotational crossing	outbreeding
	Lauprecht system	mating likes
	Gene pool	mating unlikes

Species crossing

Species crossing has rarely been exploited in livestock breeding. There are difficulties using species with different chromosome numbers since embryo survival is usually low and even if there is survival there is often sterility. Some crosses of zoological interest have been undertaken such as lion × tiger (liger or tigon) and wolf × domestic dog, which latter has little to recommend it and much to object against. There are few farm livestock species crosses undertaken.

The most famous is the horse × donkey giving rise to the mule. There have been crosses of cattle and buffalo, largely undertaken in the USA and Canada, called beefalo or cattalo; crossing of zebu and yak has been undertaken in the Himalayas. However, all of these have been of minimal interest beyond that of the mule which has been exploited at various stages in history.

Breed structure

Before discussing crossbreeding or purebreeding, some explanation of breed structure is needed. Breeds are of relatively recent origin, with the Shorthorn herd book, started in 1822, being the first such publication. Pedigree breeds, as such, can only legitimately exist as far back as their pedigrees go.

The definition of what constitutes a breed is variable but generally refers to a group of animals within a species which have certain recognisable characteristics in common which render individuals of the breed recognisable as such. Most of these characteristics are physical ones connected with coat colours and the like, such that at post-slaughter stage with the hide removed, breed identification is much harder. Many breeds have arisen because physical isolation has left the population separated from other populations and hence selection and random drift have caused distinctions to appear. Other breeds have evolved by the crossing among older established breeds until new types have emerged.

The dog is the best example of an animal species which has been subdivided into numerous remarkably different breeds. The Chihuahua and the St Bernard belong to the same species and could be mated (albeit with some practical difficulty!) to provide viable offspring. No other species can claim such physical extremes. Although canine pedigrees can only be traced back at best to 1873 – when the British Kennel Club (KC) was formed – some breeds of dog have existed in something approaching their present form for centuries but without pedigree proof. Other breeds, particularly many gun dog breeds, only trace back about 100 years.

Breeds of cat are not as distinct as dog breeds and are largely classified according to coat type, body proportions and head shape as well as coat colour. This degree of laxity is demonstrated by the introduction of the Himalayan colour pattern into the British Short-haired population. This was achieved by the use of longhaired Persians. The long coat is recessive to short coat so the colour pattern was incorporated easily into the short-coated breed, but individuals with Persian ancestors close up in their pedigree were still accepted as 'purebred' shorthairs when in most other species they would have been regarded as crossbreds. Cat breeds are each identified by a number and it is common to find pedigrees of a particular breed (duly registered as pedigree) which contain ancestors close up of other breeds. Some breed crosses would not be acceptable while others are tolerated.

Once a breed was established, breed societies or breed clubs were usually set up. These were organisations operating to foster the interests of a particular breed. Evidence suggests that all breeds, regardless of species, tend to develop a triangular-like structure, usually termed a *breeding pyramid*. At the apex would be found a small number of 'elite' herds/flocks which would be self-contained or would interchange stock with each other. These elite units, sometimes called studs or *nucleus herds*, generally sell stock to a larger group below them termed *multipliers*. These, in turn, sell to the commercial producers who may use purebred stock but which may not be registered with the society. Essentially, the movement is downwards and the pedigree registration barrier between multipliers and many commercial units prevents any upward flow at this level. Sometimes herd/flock books are 'open'. This means that breeders with other breeds can, by a series of backcrosses (usually four or five) with sires from the breed aspired to, enter upgraded stock in the pedigree register. The pattern of breed structure is shown in Figure 8.1.

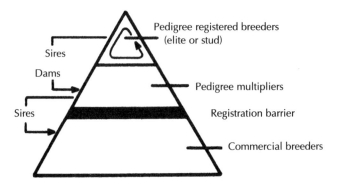

Fig. 8.1   Traditional breed structure

When AI was first established in cattle, it was considered that by using (via AI) sires which originated in the elite group, genetic merit could quickly be passed to the commercial breeders. However, it was soon found that the so-called elite units were not necessarily genetically superior to the others in the breeding pyramid. They tended to perform at a higher level for managerial reasons, but this does not imply genetic superiority. Early investigations into the British Friesian breed, for example, showed that the elite group consisted of herds which had obtained stock from the 1936 importation from The Netherlands, regardless of the merit of that stock. After some years of AI use, no improvement occurred and the nature of breeding pyramids in pedigree breeds was exposed.

The problem with a breed structure such as depicted in Figure 8.1 is that there is no upward flow of better genes. Elite herds may have excellent phenotype, but unless this is matched by superior genetics nothing is gained. Very often elite units are too small to make much genetic progress and the registration barrier is an artificial one concerned with pedigree not performance. Nevertheless, testing for genetic merit is expensive and there is something to recommend a system of testing on a smallish scale. If an elite group were to be selected on performance, then their merit could filter down through the pyramid. This is illustrated in Figure 8.2.

Breed societies/clubs are part of the agricultural tradition in many developed countries. Generally they have taken on an administrative and moral responsibility for the breed in question. They act to keep records of pedigrees and seek to ensure breed purity and

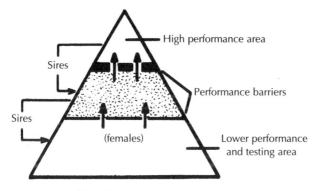

Fig. 8.2 An improved breed structure

they have served as a focus for breed promotion through a variety of activities as well as a kind of social club for those interested in the breed.

On the debit side, pedigrees devoid of performance data are of minimal value and breed societies can be inherently conservative about change. Ruling committees of breed societies were often made up of the older-established breeders unwilling to alter their views. In some cases AI was opposed, as it still is in Thoroughbred racehorses, and developments along genetic lines were often decried.

Horse breeding is controlled by breed societies or societies concerned with a particular use of the horse (e.g. show jumping or sport horses) while Thoroughbred racing is controlled in Britain by the Jockey Club. With dogs most countries have a Kennel Club-type body which controls all breeds and breed clubs and often registrations, though in Europe many breed clubs control their own registrations and pedigrees. The Kennel Club in Britain is a non-elected body which controls the activities of thousands of breeders, and though dogs have probably changed more in the past 50 to 60 years than in many centuries previously, much of the change can be seen as haphazard and sometimes even ill-directed. Those directing and controlling matters at KC level have not always possessed the knowledge or farsightedness to bring about advances, so that progress in some breeds has been almost in spite of, rather than because of, Kennel Clubs. Cats, controlled by the General Council of the Cat Fancy (GCCF), are blessed with a somewhat more democratic and progressively-minded body.

In recent years the agricultural breed societies have become more disposed towards genetic techniques and have encouraged and supported them. The concept of pedigree and breed has long disappeared from the poultry industry, aside from a few exhibition breeds of minimal influence. In similar fashion, breed is becoming less crucial in the pig industry. In Britain, the advent of new beef breeds from Europe, starting with the Charolais in 1960, served as a spur to breed societies to become more progressive and the acceptance of new ideas is quickening all the time.

Livestock shows, at one time based solely on physical appearance and often exaggerated at that, have been changing, albeit slowly. Shows provide a meeting place for breeders with common interests and an element of competition that should spur some to better things,

as well as a social occasion. Increasingly, traits other than conformation are considered, though the confounding influences of herd/
flock of origin cannot easily be overcome. Many societies now openly
encourage collaboration on scientific evaluation in a variety of ways,
which has breathed new life into previously moribund organisations.
At the same time it has led to greater understanding of animal
breeding principles and performance advances among the animal
breeding community.

In canine matters some countries have welcomed science. In
Sweden, for example, where only one breed club exists per breed and
the Swedish Kennel club (SKC) is a progressive body, much good
work has been done to combat genetic disease, and testing for
character (behavioural traits) is encouraged. The SKC actually
employs a professional geneticist. In Britain a plethora of breed clubs
for a single breed does not always make for uniformity of purpose,
though the concept of Breed Councils (collections of clubs supporting
a breed) has offset this to some degree. Nonetheless, combating
genetic disease is still an optional issue and the KC does little to
compel adherence to any code. Dog and cat breeding is carried on in
small units and thus would benefit from organised selection programmes and collaboration. Most breeders would be responsive to
rules laid down by their club (in which they participate) but are less
tolerant of rules imposed upon them and over which they have had no
say.

Purebreeding

Purebreed or pedigree breeders are, by definition, going to use
individuals of the same breed. They can either use related (*inbred/
linebred*) or unrelated (*outbred*) animals for that purpose.

Inbreeding is defined as the mating of individuals more closely
related than the average of the population from which they come.
Linebreeding is no different from inbreeding but is usually undertaken to a lesser degree of intensity, with the common ancestor(s)
somewhat further back in the pedigree than the grandparent generation.

In the establishment of breeds, inbreeding was frequently undertaken, often at high levels. That individuals of a breed are all related

in some way is obvious from the nature of a pedigree. With two parents, four grandparents, etc., there are by generation 20 over one million ancestors in that generation alone. In cattle, 20 generations could be about 100 years and that long ago there were probably not a million members of the breed. Clearly many ancestors must be duplicated and hence relatives have been mated together. For this reason the definition of inbreeding refers to 'more closely related than the average'.

Inbreeding is measured by the inbreeding coefficient (F) devised by Sewell Wright in the 1920s. An illustration of how to calculate this is shown in Appendix 1. As a system, inbreeding arouses a great deal of anxiety among lay persons and some practical breeders. In humans there are limits to the closeness of mating, with first cousins being the closest mating permitted. This would give rise to progeny with an F value of 6.25%. These controls on humans have not always existed and sibling matings were common among some ancient civilisations. There are no such limits on animal breeding and problems can arise.

The first and often most obvious consequence of inbreeding relates to deleterious genes. Inbreeding does not create defects, but, because many defects tend to be recessive in nature, the mating of close relatives increases the chances of such deleterious genes coming together and giving rise to abnormal progeny. Since all individuals will carry a number of deleterious genes, the occurrence of abnormalities is inevitable with inbreeding. Undesirable though this may be it is not a major disaster.

Defects have to be assessed in relation to their seriousness and the qualities of the animal which produced them as well as to their age at onset. If a dairy bull is an outstanding improver of milkfat/protein yield it would be ludicrous to cull him simply because once in every 10 000 calvings he produces a defect like 'amputate' which is a dead calf born with amputated feet and a long, parrot-like muzzle. The loss of a calf, even if more frequent that one in 10 000, is of minimal economic importance relative to the advantages that the bull can give. Those who become emotionally involved over such issues must realise that all stock carry defects and there is currently no way they can all be eliminated without ceasing to breed altogether. In humans the situation would be different, but we cannot use human arguments in a livestock situation.

Of greater importance than the occasional defective offspring is the consequence of inbreeding depression. This is the gradual lowering of performance with increasing inbreeding. Usually, the traits most affected will be those concerned with fitness which tend to be traits that are of low heritability but highly influenced by non-additive genetic features, especially dominance. As inbreeding progresses there will be a decline in fertility, an increase in embryonic mortality, a decline in progeny survival and eventually a lowering of growth rate and milk yield, though highly heritable traits such as some carcass features might be unaffected. It can be shown that inbreeding depression is measured by the formula:

$$M_F = M_0 - 2Fdpq$$

where M_F is the population mean after inbreeding, M_0 is the population mean before inbreeding began, F is the inbreeding coefficient, d is the degree of dominance and p and q are the gene frequencies as described in Chapter 4 (see 'Gene mapping'). The definition of dominance can be demonstrated using the gene symbols A and a. We have three genotypes AA, Aa and aa and if there is no dominance then Aa will appear intermediate between the phenotypic values of AA and Aa. The extent to which Aa is above the mid-point between AA and Aa is a measure of dominance or d.

If AA measured 20 units and aa 10 then the mid-point is 20 + 10/ 2 = 15. If Aa measured 18 units then the degree of dominance is 18 − 15 = 3 units.

In situations where dominance is minimal there will be less effect of inbreeding, but the greater the degree of dominance the greater the potential depression that may be seen. This reinforces the view that highly heritable traits will tend to be less harmed by inbreeding than those with low heritabilities and high dominance influences. Of course the greater the inbreeding (F) the greater the risk. One might expect few problems up to a level of say 10% inbreeding but increasing difficulties in excess of 20%, though this will depend upon the species and population being studied. It may be that the speed with which inbreeding is achieved may also be more important than the level of inbreeding. Some individual populations can withstand high levels of inbreeding with no apparent ill effects, but for every such success there will be many more failures. The Chillingham Wild White cattle herd in Northern England is an example of a closed population running for three centuries and still surviving despite its inbreeding.

Lamberson and Thomas (1984) published a review of inbreeding effects on sheep based on 25 studies covering some 25 000 sheep. A summary is shown in Table 8.2. In most traits, inbreeding has an adverse effect, with the greatest harm being to lamb survival.

Table 8.2   Effects of inbreeding in sheep

Trait	Numbers of observations	Regression coefficient	
		Individual	Dam
Greasy fleece weight	>6356	−0.017 kg	0.005 kg
Clean fleece weight	>8054	−0.001 kg	
Staple length	>13612	−0.008 cm	0.002 cm
Birth weight	3678	−0.013 kg	−0.013 kg
Weaning weight	10183	−0.111 kg	−0.072 kg
Post-weaning weight	>5777	−0.178 kg	0.013 kg
Body type score	9902	0.011	0.010
Condition score	9902	0.012	0.017
Face cover score	>11902	0.002	0.014
Wrinkle score	>12630	−0.020	0.002
Lamb survival	>6266	−0.028	

Regression shows change in performance for each 1% increase in inbreeding of the lamb or of the dam
Source: Lamberson and Thomas (1984)

Linebreeding is a term used when the breeder tries to combine lines which trace back to a favoured ancestor. The pedigree used in Figure 6.1 could be said to be linebred on Champion of England. In reality it is an issue of degree. The consequences of inbreeding are also those of linebreeding, but because the F value may be less with linebreeding, adverse consequences may also be less. The levels of inbreeding which will occur in a herd/flock that is closed to outside materials and uses varying policies is shown in Table 8.3.

This table shows how very close matings (e.g. full- or half-sib matings) lead to a rapid build-up of inbreeding whereas with less close relationships, even with relatively few sires annually, inbreeding does not build up to the same degree. For the 1-, 3- and 5-sire herds/flocks, a generation interval of 5 years is assumed. Thus, 1, 3 or 5 new

Table 8.3 Extent of inbreeding (%) for different closed herd/flock programmes

Generation	Offspring/ parent or full-sibs	Half-sibs	One-, three- and five-sire herds/flocks New sires per year (5 year generation)		
1	25.0	12.5	2.5	0.8	0.5
2	37.5	21.9	5.0	1.6	1.0
3	50.0	30.5	7.5	2.4	1.5
4	59.4	38.1	10.0	3.2	2.0
5	67.2	44.9	12.5	4.0	2.5

sires per year equates to 5, 15 or 25 new sires per generation. It is also assumed that mating is at random in these units. By careful attempts to reduce close matings the build-up would be less than that given.

Inbreeding increases prepotency which is defined as the ability of an individual to stamp its virtues (or deficiencies) upon its progeny. Although used in the development of all breeds, most farm livestock species are now bred with some emphasis upon reducing inbreeding or keeping it at a minimum. Mostly this arises from the knowledge of the damage that can be done to fitness traits. However, in cats and dogs where litter size would be less important than individual excellence, inbreeding may be useful to breeders with outstanding stock prepared to sacrifice something in terms of numbers of pups/kittens in order to advance with conformational/functional issues.

Inbreeding levels are much less in most breeds, especially numerically large ones, than many lay people imagine. Inbreeding can be divided into current inbreeding (i.e. within the first two generations and therefore something which the breeder is fully aware of and may well have undertaken deliberately) and non-current inbreeding. The latter can be subdivided into long-term inbreeding and inbreeding caused by the separation of the breed into strains. Breeders tend to work with relatively short pedigrees (four or five generations) and thus may be unaware of more distant relationships. Even when breeders are making no attempt to inbreed, they will be using individuals that show some relationship because the breed is a closed unit (albeit a large one). In many livestock breeds inbreeding may be advancing at around 0.33% per year.

Schemes set up to advance breeds cannot always be large enough in terms of animal numbers to avoid inbreeding. It is possible for inbreeding depression to be large enough to offset selectional advances. This is a problem in the perpetuation and preservation of rare breeds or species. Because they are, by definition, limited in numbers, inbreeding can arise and can be sufficient to reduce fertility and thus make the preservation of the species/breed even harder.

Outbreeding is the mating of animals of the same breed but with minimal relationship between them. Exactly how unrelated is a matter of opinion, but calculation of the coefficient of relationship is shown in Appendix 2.

When breeders mate animals with the same virtues it is termed *like-to-like* or *assortative mating*. When this is undertaken it can have a bearing upon the evaluation of sires. Most breeders interested in conformational aspects will undertake like-to-like mating at some stage and it is a common policy in dog/cat breeding. It should, however, be realised that animals which may look alike do not necessarily carry the same genes.

Compensatory mating or the mating of *unlike to unlike* is also practised by breeders. Again, this is much followed by dog/cat breeders. At some stage a particular conformational/behavioural failing may appear to be prevalent in a line and the breeder may seek to rectify this by using animals that do not possess these failings but actually excel in these features. This policy tends to make the population more alike.

Crossbreeding

The basis of crossbreeding is the exploitation of *hybrid vigour* or *heterosis*, the latter term being preferable. If two breeds or lines are crossed, then one can expect the progeny to be at the mid-point between the parental means. If the progeny exceed the mean of the two parents then the extent to which they exceed the mid-point is a measure of heterosis. In plants, heterosis is usually confined to progeny which exceed the best parent, but in animals this is rare. Crossing a numerically small but superior breed with a numerically large but inferior breed may give progeny that exceed the mid-point but not the better parent. This does not diminish their value since, in one generation, the numerically large but inferior breed is improved much further than might have been feasible by direct selection.

Traits that are depressed by inbreeding are usually the ones that are enhanced by crossbreeding. Heterosis in the first cross can be shown to be equal to:

$$H_{F1} = dy^2$$

where H_{F1} is the heterosis in the F1 generation, d is the degree of dominance (as explained in 'Purebreeding') and y^2 is the square of the difference of the gene frequency between the two parental lines/breeds. This suggests that to a great degree heterosis is the reverse of inbreeding depression. Traits which show no dominance should exhibit neither inbreeding depression nor heterosis. The formula above relates to a single locus and combined over all loci it becomes:

$$H_{F1} = \sum dy^2$$

Heterosis declines in the F2 generation to half that in the F1. Greatest heterosis is likely when the breeds crossed differ markedly in their gene frequencies and the trait(s) under consideration is under dominance control. Heterotic effects are often high when crossing temperate cattle (*Bos taurus*) with tropical ones (*Bos indicus*). Some estimates of heterosis are shown in Table 8.4.

Grading-up

Grading-up is the gradual changing of a breed by continued crossing in each generation by a second breed. The system has been used all over the world to 'upgrade' one breed by converting it into another. Most breed societies with open herd/flock books will accept four or five topcrosses of this kind and still consider the upgraded animal purebred. After four such crosses the animal would be 93.8% pure. The better the sires used in such a programme the better the end product, but if the new breed was considerably superior to the existing one, even the use of random sires would result in progress. Because heterotic effects decline in the F2 generation, grading-up does not allow exploitation of heterosis much beyond the initial cross. The virtual demise of the Shorthorn in Britain arose by grading-up to

Table 8.4   Some general estimates (%) of heterosis

Species	Trait	Heterosis (%)
Dairy cattle	Milk yield	2–10
	Milkfat yield	3–15
	Feed efficiency	3–8
Beef cattle	Calving rate	7–20
	Calf viability	3–10
	Calves weaned/cow mated	10–25
	Birth weight	2–10
	Weaning weight	5–15
	Feedlot gain	4–10
	Carcass traits	0–5
Sheep	Barrenness	18
	Lambs born/ewe lambing	19–20
	Lambs weaned/ewe mated	60
	Birth weight	6
	Pre-weaning growth	5–7
	Carcass weight	10
	Fleece weight	10
Pigs	Piglets/sow farrowing	2–5
	Litter size/weaning	5–8
	Litter weight/weaning	10–12
	Growth to slaughter	10
	Carcass traits	0–5

After Dalton (1985)

Friesian in the same way that the British Friesian is currently well on the way to being graded-up to Holstein.

Backcrossing/crisscrossing

A backcross occurs when a first cross animal (F1) is mated back to one of the original parent breeds. This is a common policy with pigs in Britain using Large White and Landrace and mating the F1 of these two to either the Large White or Landrace. Although the resultant slaughter generation is 75% of one breed and 25% of the other, the scheme does allow exploitation of heterotic effects from the use of a crossbred dam.

Crisscrossing is an extension of backcrossing. Initially, two breeds are crossed (A × B). The crossbred is then mated back to A to give 75% A and 25% B. This crossbred is then mated back to B to give 62.5% B and 37.5% A. Thereafter breeds A and B are used alternately. This policy was once widely used with American breeds of pig.

Rotational crossing

In rotational crossing, a series of breeds (three, four or more, since two constitutes crisscrossing) are used in succession. This is shown diagrammatically in Figure 8.3. In principle, by using several breeds, there ought to be greater heterosis than by using a crisscrossing policy. Studies in cattle and pigs have shown that for several traits three breed rotational crosses can be superior to purebreds by up to 25%. Much will depend upon the trait being studied and the breeds involved in the programme, but it does appear that the rotational system has much to recommend it for crossbreeding.

The Lauprecht system

The Lauprecht system was first put forward by a German scientist (Lauprecht, 1961) and, essentially, it consists of a crossing pro-

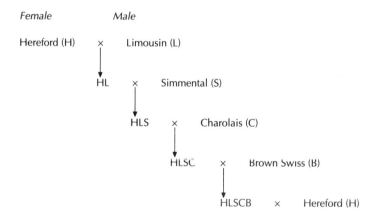

Fig. 8.3 Rotational crossing using five breeds

gramme using three breeds (A, B and C). A is crossed with B and the resultant progeny crossed with C. Thereafter, the policy is somewhat different from other programmes because only crossbred males are used in each case, with the female generation being mated to males from the previous crossbred generation (see Figure 8.4).

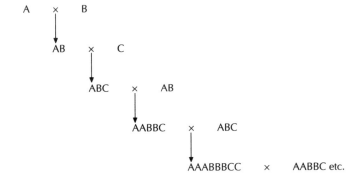

Fig. 8.4   The Lauprecht system

Essentially, the system results in a population which, by about the fourth cross and all crosses thereafter, is approximately one-third of each breed. Although this is an unusual programme, the disadvantage is that it effectively becomes a new breed rather than a distinct crossbred. Whether it has been used successfully in a practical context is uncertain.

Gene pool

In the gene pool system, a number of breeds are used as a combined population with selection for specific traits, selecting animals solely on merit rather than on cross or breed. New breeds can be added but only if they have performance criteria to justify this and not simply because they are new. In the 1960s the Animal Breeding Research Organisation (now the Roslin Institute) developed a low backfat line of pigs based originally on four breeds. The line was open to immigrants from any promising source and showed a relatively large decline (improvement) in backfat over the first 7 years compared with breeds in progeny test stations.

Creating new breeds

Although most breeds in a species trace back to the nineteenth century or earlier, new breeds are still being formed. Sometimes breeds are little more than a cross between two foundation breeds with interbreeding among F1 animals. The New Zealand Half-Bred sheep is a case in point being a cross between the English Leicester and the Merino. Any expected decline in the F2 needs to be offset by selection within the population.

Other new breeds have been developed for specific situations, which is especially true in cattle where attempts to find a combination of *Bos indicus* and *Bos taurus* suitable for the tropics have been much practised. Often there has been a deliberate attempt to use about five-eighths *Bos taurus* and three-eighths *Bos indicus* though there is nothing 'magical' about this specific combination. Examples (not all five-eighths and three-eighths) with area of origin are shown in Table 8.5.

The way in which the Siboney de Cuba was produced is shown in Figure 8.5.

Table 8.5 'Newly' created livestock breeds

Breed	Sp.	Components	Location
Santa Gertrudis	Cattle	Shorthorn/Brahman	USA
Luing	Cattle	Shorthorn/Highland	Scotland
Australian Milking Zebu	Cattle	Jersey/Zebu	Australia
Jamaica Hope	Cattle	Jersey/Sahiwal/Holstein	Jamaica
Brangus	Cattle	Angus/Brahman	USA
Charbray	Cattle	Charolais/Brahman	USA
Siboney de Cuba	Cattle	Holstein/Zebu	Cuba
Beefmaster	Cattle	Shorthorn/Hereford/ Brahman	USA
Droughtmaster	Cattle	Shorthorn/Brahman	Australia
Colbred	Sheep	East Friesland/ Clun Forest/ Border Leicester/ Dorset Horn	England
Finn–Dorset	Sheep	Finnish Landrace/ Dorset Horn	England

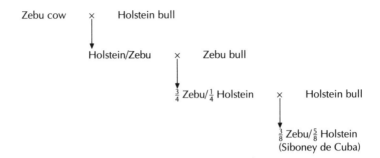

Fig. 8.5   Method of producing the tropical milking breed – the Siboney de Cuba

Exactly how successful all these 'new breeds' are is open to question. The idea of producing a new breed for specific environmental conditions may seem highly attractive, but such programmes may take 20 years or more. By the time a breed has been developed for 'today's conditions' it is actually 'tomorrow' and those conditions may have been changed by better nutrition and management. Frequently, new breeds are gimmicky and greater progress might have resulted from the same investment in management improvements.

Chapter 9
Breeding in Practical Terms

General principles

Each breeder is a distinct entity, faced with different circumstances in terms of species/breed, herd/flock size, location, management techniques and objectives. It is not therefore possible to lay down hard and fast rules for any given set of circumstances. This chapter looks instead at broad overviews.

The decision as to which traits to select for will be determined by their relevance to the overall economic objective, rather than simply picking something because it is highly heritable or easy to measure. Ponzoni and Newman (1989) examined this in relation to beef cattle, but the principles have much wider application. A breeder has to distinguish between his breeding objectives and his selection criteria. Those traits in the breeding objective may be considered the ends while the characters used in the selection criteria are the means to achieving those ends.

Let us take a Suffolk breeder who is breeding pedigree stock. He may sell some animals directly for slaughter, but, in the main, he is in the business of selling breeding stock, principally rams, to those who either are pedigree breeders like himself or want terminal sires to cross with another breed. He could be quite successful in achieving high ram prices if he was to select for large heads and strong legs, which might be termed conformational aspects. These features may, alas, have little relation with genetic superiority in terms of growth rate and meat production, but they may be highly sought after by some Suffolk breeders who would pay well for rams with these virtues. If, instead, he decides that his long-term market must be to produce good terminal sire rams then he must be concerned with growth rate, more particularly lean growth rate and with muscularity increases and fat level decreases. He would probably be best selecting for an index incorporating these features. The prices he will obtain for his rams will be related to index level.

He might also be interested in reducing the risks of scrapie. This would be advantageous within his own flock and would command additional financial benefits on sold stock. DNA typing can allow the identification of RR, RQ or QQ Suffolks, with RR animals being resistant to scrapie, RQ animals being largely resistant and QQ animals susceptible. Blood sampling and DNA testing would allow the flock to be identified as to scrapie risk and selection could be a two-stage one: culling QQ animals and eventually RQ animals while selecting on index terms among RR animals.

This is just a rough assessment. In drawing up an objective the breeder must look at the enterprise as a whole and take account of all the inputs and outputs, particularly as this affects income and expenditure. Traits selected must be based upon their relevance to the overall economic objectives. If a high cost is feed then feed efficiency clearly has economic merit as a selection trait. If animals are sold as carcasses then those aspects which contribute to carcass value such as leanness and muscularity will have merit in a selection programme.

It has been shown earlier (see Chapter 5) that progress depends upon the heritability of the trait, the selection differential and the generation interval, all of which can, to a greater or lesser degree, be influenced by the breeder. It has also been shown that the routes to progress are:

(1)   Selection of males to breed males.
(2)   Selection of females to breed males.
(3)   Selection of males to breed females.
(4)   Selection of females to breed females.

In general terms route (1) is the one with the greatest opportunity for progress, closely followed by route (2). Route (4) is the least useful unless embryo transfer techniques are in operation.

Most breeders are in control of the female side of things since they will have a herd/flock or they will possess females and determine which females are used for breeding and how they will be mated. Males may also be owned by the breeder and this will be true of sheep and pigs but less so of cattle, horses and companion animals. Indeed, a small breeder of companion animals might be best advised to fill his/her available space with females since, for the right stud fee, access is possible to the best available males in the country. A stud male will earn income, but keeping more females and having access to the best males that can be located would be the wisest use of space. This is least true in cats where studs which are not open (available to all) are commonplace and thus access to the best males may not be possible unless one has them in one's own cattery.

Increasingly, large commercial companies have become involved in the animal breeding world. This has long been true of poultry and is increasingly true of the pig industry. Many people who might once have bred their own males and females and thus had some control over their breeding programmes are now buying stock from breeding companies. If a pig producer buys gilts from a pig breeding company and uses bought-in semen or purchased males to use on them, with a view to sending all progeny for slaughter as meat animals then the producer is no longer a true breeder. The only breeding decision being made is which company to go to for stock and the producer simply farms those stock and replaces them with more purchases when the time comes. To this degree the breeding decisions are increasingly in the hands of breeding companies and not the one-time breeder. As yet, sheep are still largely bred in traditional fashion with limited use of AI and no large companies, certainly not in Britain. However, although dairy cattle are still within the breeding control of dairy farmers, the breeding of dairy sires is moving more towards companies like Genus in Britain and similar companies in other locations.

Breeding companies can operate on a larger scale than most individual breeders and have access to more sophisticated computing techniques as well as affording genetic expertise. As such, breeding may advance more rapidly than under more traditional systems. On the other hand, commercial companies may be reluctant to publish much about their methods and equally reluctant to collaborate in comparative evaluations with other companies. Commercial companies may make decisions that are best for their businesses but not necessarily for individual livestock producers and the choice of where to go for stock will become crucial. In the past, quasi-governmental organisations such as the Meat and Livestock Commission (MLC) or the, now defunct, Milk Marketing Board (MMB) provided valuable data for all and sundry which is increasingly hard to obtain. The MLC still operates but with less ability to create data.

Identification

In any programme, accurate identification of animals is important and this cannot be overemphasised. Lost identification or misread tags can cause havoc with breeding programmes and result in the loss of valuable and expensive data. By the same token, the use of false pedigrees whether by accidental misidentification or by deliberate fraud is highly damaging to breeding programmes.

Permanent identification is not always as permanent as it sounds, but could include:

- metal ear tags
- plastic ear tags
- ear tattoos
- freeze brands

- fire brands
- plastic tail, hock or ankle tags
- ear notching
- electronic implants.

Not all of these systems apply to all species. Dogs are usually identified (if at all) by ear tattoos or by implants; cats can be tattooed but are probably best identified by implants. Ear notching is used in pigs and to a lesser degree in sheep but is likely to run, increasingly, into 'welfare criticisms'. Ear tags are useful in farm livestock but can be lost, particularly plastic ones, though these have the advantage of being available in numerous colours. Tags are not readable from a distance and metal tags require the animal to be restrained to be read. Ear tattoos also require restraint and can become illegible as well as being dependent upon ear colour, but they have been a system of choice in canine identification (where restraint is not a problem) with considerable success. Fire branding is largely a beef ranching system which, though fairly effective, damages the hide. The same is true of freeze branding, but brands show up as white hair and are best on dark coats. Dairy cows are often freeze branded on their rear for ease of reading in the milking parlour. Plastic tags on the tail, hock or ankle are dairy cattle features also chosen for ease of reading during milking.

Implants, usually inserted in the neck region, have been expensive and require an even more expensive instrument to be able to read them. However, both the implant and the transponder are getting cheaper and if there is no migration of the implant within the body this system can allow identification at the carcass stage whereas all other identification procedures tend to be lost from this point. Increasingly, farm livestock breeders require identification through to the carcass stage. DNA fingerprinting can, of course, identify animals, but it is not something that can be used instantaneously. Various organisations now exist which will DNA test parentage.

Semi-permanent identification systems include:

- neck tags of wood/plastic attached by a cord
- hair dyes or bleach used on the animal's side
- paint brands
- plastic-covered wire twisted into a hole in the ear.

Hair dyes or bleach numbers on the side (usually in sheep) can be read at considerable distance, as can paints, but these should be used sparingly and only approved washable paints used. By definition, semi-permanent systems are intended for relatively short periods of time and are easy and useful. Thus, identification (after ultrasound) of ewes bearing singles, twins or triplets can be done using different coloured paints and be very effective for the short time that it is needed.

Very temporary markings would include stick-on labels and washable paint sprays or chalk raddle marks. They apply only for very short periods. It might be useful to have more than one system in operation. For example, metal tags are frequently used on calves but are not easily read. Large plastic tags in different colours (to indicate sire, group, dam age, etc.,) can be used alongside and, if lost, the metal tag is still available. The more animals that there are in a programme and the more extensive the management system, the greater the risk of lost identification and hence inaccurate or lost data.

Dairy cattle breeding

Dairy cattle breeding is usually directed towards milk yield, milk composition or quality, longevity and some aspects of conformation.

All developed countries and many developing ones have official milk recording systems which should be consulted. The more frequently milk is measured the greater the accuracy of that measurement, but weekly or monthly recording is accurate enough for many purposes. Testing for compositional quality requires laboratory facilities, but machines now exist which can evaluate numerous samples very quickly.

Milk yield, milkfat and protein yield are of moderate heritability (ca. 30%) while the heritability of milkfat and protein percentage tends to be high (ca. 55%). Regularity of calving and longevity are of low heritability while conformational aspects vary. The fact that milk is assessed only in females complicates selection procedures and generally requires bulls to be progeny tested. Correlations between milk and quality traits tend to be positive and high when in weight terms, but milk yield and fat percentage are slightly negatively related.

Most dairy farmers will undertake selection for such traits as temperament and cull those cows which are excessively nervous or slow milkers. Usually such animals would not yield well in any event because they will rarely be given the excessive time required for complete let down of milk. Culling also takes place on the grounds of disease (e.g. susceptibility to mastitis) as well as on poor reproductive performance. In both of these cases there may be some association with yield, as high yielders may be more prone to mastitis and reproductive problems. This is unfortunate, but no herd has a place for cows of poor fertility or high disease risk. It does, of course, need to be remembered that failure to rebreed may in some instances be a failure of stockmen to accurately identify cows in oestrus.

It is useful to record heat periods, services and calving dates to check on fertility, and, in addition, some record of calving difficulty is useful. This will be subjective and usually is done on a scoring system such as:

0 = no assistance
1 = slight assistance
2 = veterinary assistance
3 = considerable assistance using mechanical/veterinary aids
4 = caesarian section.

Birth weights of calves might be recorded along with gestation length, if known (likely if AI was used), as well as the sex of calves. Milk would be recorded by weight or volume and composition in absolute and percentage terms. The lactation length is needed as well as the number of times milked per day. A three-times per day milking might be expected to add 10% to milk yield. One cannot compare twice- or thrice-daily milkings without applying correction factors. Although total lactation yield may be recorded most countries consider the official lactation yield to be 305 days or less (if the cow dried off earlier) with lactations of less than 100 days not being considered official. When using 305-day yields, records are truncated at that point if the cow milks longer than 305 days.

Some aspects of type would be valuable, if these can be related to practical features such as disease resistance or longevity. Unfortunately, longevity is not easy to measure and has the practical problem of taking a long time to assess.

Generally, the best cows in a herd would be bred to bulls of the same breed and the poorer cows to beef bulls if there is a need for beef from the dairy herd. Culling on yield would be undertaken at the end of each lactation but only after culling on fertility, disease, etc. has been done.

Methods of assessment

Dairy cow selection began around the 1930s, though some selection will have occurred much earlier. Initially, selection was on actual yield, but it was quickly realised that this merely reflected managerial differences. The next stage was to assess daughter–dam comparisons. Here again there was a problem in that daughters and dams were milked in different years and, possibly, even in different herds.

Contemporary Comparisons (CCs)

The first real breakthrough came in the 1950s with the application of Contemporary Comparisons (CCs) which was made possible as computing facilities developed. The system was based upon evaluation of first lactation yields – which were more unselected than later yields. Comparisons were made between the first-lactation daughters of a specific bull under examination and the daughters of all other bulls milking in the same herd at the same time. This was done in all recorded herds and the information pooled. Data in any herd were weighted (W) according to reliability based on the formula:

$$W = \frac{\text{number of daughters} \times \text{number of contemporaries}}{\text{number of daughters} + \text{number of contemporaries}}$$

Thus, if a bull had four first lactation daughters in a herd that had a total of 12 heifers, animals milking at the time, then weighting (or what can be called the number of effective daughters) would be:

$$W = \frac{4 \times 8}{4 + 8} = \frac{32}{12} = 2.67$$

There was some criticism from breeders that selection on first-lactation yield would mean selection for cows which 'burnt out' quickly. It was subsequently shown that the high CC bulls had more daughters lasting longer in herds than did low CC bulls. The CC was given as a + or – value in kilograms and was the amount by which a bull's first lactation daughters could be expected to outperform 0 CC bulls. Although used in Britain, there were similar systems developed in other countries, all based upon comparisons within herds and years or herds and seasons, sometimes called herd–mate comparisons. Mostly first lactation yields were used, but in countries with small herd size, and thus limited numbers of first lactation animals per herd, records of all cows were used by converting lactation yields to mature equivalents (MEs) which was the record expected if the cow had been a mature animal.

Progeny testing schemes were set up in which young bulls, sired by the leading high CC sires out of high performing cows, were mated to some 300 cows and then 'laid off' while records of daughters were awaited. Once progeny data became available the better CC animals returned to service and the poorer ones were culled. This meant that a continuous improvement system was in operation, but it was realised that age at calving and month of calving had a bearing upon lactation yield. It was also realised that, as progeny testing became more accepted and greater use was made of the better bulls, young bulls were handicapped. Instead of being compared against a random or average sample of contemporaries, young bulls were increasingly being evaluated against superior progeny tested bulls.

In the 1970s, with the advent of more powerful computers, the CC system was improved to give the Improved Contemporary Comparison (ICC). Under this method, assessment was still made on first lactation daughters and records were weighted together as before, but now a correction was made for age at first calving (usually to a 28-month mean age) and for the month in which the heifer calved. Contemporaries were then those milking in the same herd/year/season, with season being a 3-month period for contemporaneity. Additionally, a correction was made for the quality of the bulls with which the young bull under assessment was being compared. ICCs were related to a base year and continued to be used for a decade or so.

New systems

By this time the easy assessment of protein content as well as milkfat was in operation and milk producers were being paid on both components. With the advent of BLUP techniques the introduction of new systems of assessment was made feasible. Genetic indices began to be used as the best estimate possible of the animal's ability to transmit its genetic merit. A genetic index would thus take account of the animal's own performance (if available), the performance (index) of the sire and dam and other close relatives and then the performance of progeny. As information accrues about the animal itself and eventually its progeny, then the value of relatives declines.

Predicted Transmitting Ability (PTA) Index

The Predicted Transmitting Ability (PTA) Index is the most commonly used index and is a genetic index for production that can be calculated for males or females. Thus, a PTA for a specific bull might read:

804 kg milk	24.3 kg protein
33.2 kg fat	–0.01% protcin
–0.06% fat	83% reliability

This would mean that the bull's daughters would be expected to give, on average, in their first lactation 804 kg more milk, 33.2 kg more fat and 24.3 kg more protein than daughters of a sire which had a PTA of zero for each of these features. The minus figures for percent fat and protein are because the bull is lowering these percentages, even though raising actual yield. Reliability is a reflection of the number of daughters involved in the test. The higher the reliability the more likely it is that the proof will be a true reflection of the bull's transmitting ability. Low reliability figures should be given little credence and values under 50% are not usually published. PTAs are related to a base year and were originally labelled as PTA90s. Later, the base year was moved forward and they became PTA95s. Comparisons should not be made between different base years and the base year will change every 5 years.

The Cow Genetic Index (CGI) was an early index giving a single figure value as to a cow's genetic merit. It was based upon an equal emphasis of kilograms of fat and of protein, but it is now rarely used. The formula was:

$$10(\text{PTA fat kg} + \text{PTA protein kg}) + 500$$

Profit Index (PIN)

The Profit Index (PIN) is an economic index derived from the PTA and is expressed in £ sterling. It reflects the expected increase in revenue per lactation per daughter relative to a bull with a PIN value of £0. The PIN formula was calculated to take account of milk price, feed energy required for each component and other economic costs. It is assessed from the PTAs of milk, fat and protein yield each multiplied by a weighting factor. Currently this is:

$$\text{PTA milk kg} \times -0.30 + \text{PTA fat kg} \times 0.60 + \text{PTA protein kg} \times 4.04$$

PIN values have replaced ICCs as the description of choice for the merit of a bull or cow in Britain.

Index of Total Economic Merit (ITEM)

The Index of Total Economic Merit (ITEM) is, like PIN, an index of economic merit but takes account of factors other than yield, including herd life or longevity. The formula operates like PIN and gives a value in £ sterling, but, in addition to the milk/fat/protein values as given under 'The Profit Index (PIN)', there are additional values based on conformation. These are:

Angularity × 1.8 + foot angle × 1.1 + udder depth × 2.7+ teat length × −2.5

These are all conformation measures thought to have a bearing on longevity. They are based upon linear assessment systems developed by the MMB from 1983 which examine 16 traits, each of which is scored from 1 to 9, with the biological mid-point being 5. Thus, stature, for example, would range from a score of 1 (animals with a wither height of 125 cm or less) through 5 (137 cm) up to 9 (149 cm or more). Selection on the ITEM would not only increase milk, fat and protein yield but also angularity, and give a steeper foot angle, a lower udder depth and longer teat length. Angularity has recently been dropped from the formula.

Table 9.1 gives the percentile levels for each trait as at July 1996. This shows that the top 1% of bulls would have had a mean PIN value of £65 while the top 20% would have averaged £25.

Table 9.1 Top percentile levels for various traits/indices as at July 1996

Top %	Milk (kg)	Fat (kg)	Protein (kg)	PIN (£)	ITEM (£)
1	587	22.1	17.1	65	64
5	408	15.9	12.0	46	45
10	319	12.7	9.5	36	35
20	218	8.8	6.6	25	24
50	40	1.4	1.2	5	3

Source: Holstein Friesian Society (1997)

Pedigree Index

The Holstein Friesian Society developed the Pedigree Index which is an estimate of the genetic potential of a young bull (or cow). It is based on PTAs of parents or information further back in the pedigree. Since it is not based on the animal's performance or that of its progeny it is less reliable than other indices described above, but it does give a guide to the potential of the bull. Figure 9.1 shows the

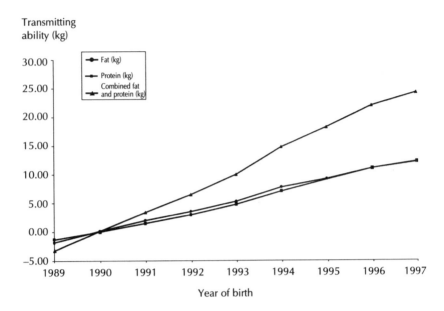

Fig. 9.1   Trends in Pedigree Index of UK females born 1989–97 (from the Holstein Friesian Society, 1997)

upward trends in female pedigree indices for fat, protein and both combined (CFP) for females born from 1989 to 1997.

Multiple Ovulation Embryo Transfer (MOET) schemes

The advent of multiple ovulation and embryo transfer (MOET) led to new ideas on the testing of bulls. Nicholas (1979) and Nicholas and Smith (1983) published ideas for MOET schemes and these are now established in different locations. They are complex but essentially involve assessing sires on the basis of pedigree information together with information on full and half sisters of the bull.

The basic principle is aimed at reducing the generation interval from some 6–7 years of traditional progeny testing to about 3 years. Progress will depend upon the number of progeny used in transfers and the methods of selection. It was originally envisaged that a scheme with about 100 embryo transfers and a milk recorded herd of 500 females would give 30% more progress compared with conventional progeny testing schemes. This was later scaled down slightly as the consequences of inbreeding were taken into account.

A MOET scheme is an attempt to produce information from relatives along the lines of litter data available in a species like the pig. One such scheme was established in Britain by Premier Breeders, later taken over by the MMB and now run by Genus. The scheme originally involved 250 cows as a nucleus unit, based upon high merit stock, many of North American origin. Some 500 embryos were planned from the best 32 cows and 8 carefully selected sires. This would lead to about 130 successful transplants of each sex. If each of the 8 sires mated 4 of the 32 cows and produced an average of 4 daughters (and 4 sons) per mating then a bull could be assessed on:

His dam's record + 4 full sister records + 12 half sister records.

A heifer could be assessed on:

Her own record + her dam's record + 3 full sister records + 12 half sister records.

A plan of the original format is shown in Figure 9.2. This has been modified by Genus and only the top 25% of cows are now bred from.

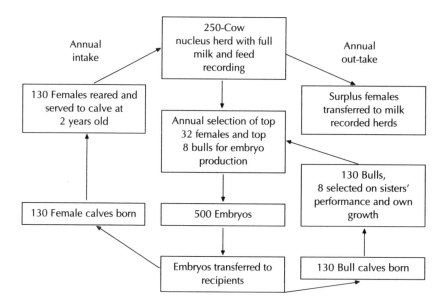

Fig. 9.2 Original plan for a MOET nucleus herd

The herd is an open nucleus which means that suitable females from other locations can be introduced, which all helps to minimise inbreeding. Bulls tend to be produced in groups of four and the predictions of merit will apply to all four though not all four would necessarily be used. Because bulls are selected essentially on sibling records, generation interval is about 2.5 to 3 years. Cows do not remain long in the Genus herd but can be sold as high index animals to other breeders so that breeding life is not curtailed. Because bulls produced in this herd are used at large, progeny test data will accumulate and allow comparisons to be made between progeny and predicted tests. Thus far these tend to show high agreement in the majority of cases.

There are some political difficulties in that, having convinced farmers of the value of using progeny tested sires, they may be less willing to use bulls on the basis of sibling tests which – though quicker – are less accurate. However, the fact that progeny test follow-up data are available can offset this to a degree. An advantage of the MOET scheme at Genus is that by keeping cows indoors all the year round on specific diets it is possible to record information such as feed intake and milk let-down – features not readily assessable in commercial herds. The risks of having a national breeding unit struck down by diseases like foot and mouth are minimised by having the milking herd and young stock in different locations.

The future

Increasingly, dairy cattle breeding cuts through national boundaries so that sires in many locations are available for use and conversion factors exist to allow sire indices in one country to be expressed in the systems of other countries. Unlike other farm species, where crossbreeding is frequently followed, dairy cattle breeding is increasingly geared towards the black-and-white Holstein-Friesian animal, with other breeds becoming increasingly rarer. Largely this results from the sheer superiority of the black-and-white animal relative to other breeds, especially when features relating to beef are also considered.

Beef cattle breeding

The end product of a beef enterprise is a carcass, but many beef producers might consider the weaned calf to be their end product.

The enterprise is, by nature, more extensive than dairying in that beef cows are not 'seen' on the daily basis that dairy cows are and AI is less often applied.

Beef breeders are interested in reproductive performance, ease of calving and the milking ability of their cows coupled with growth potential and lean meat production in the progeny of those cows. The traits concerned with growth and leanness are measurable in both sexes. Beef cows are increasingly crossbred in order to allow the exploitation of heterosis with valuable effects upon reproductive performance. Calving difficulty is subjectively assessed as described for dairy cows (see 'Dairy cattle breeding'), but milking performance is usually assessed by looking at calf growth. Many authorities suggest that small-to-medium-sized cows are preferable to large ones because the former have lower maintenance requirements and dystocia is less with smaller cows.

Calves are usually tagged and recorded within 24 hours of birth, which includes taking birth weight within this period. Weighings can be taken at intervals but 100- and 200-day weights are frequently taken. Weaning age is somewhere around the 200-day mark. However, calves tend to be weaned on a fixed day but are born over a period of several months which means that correction to a fixed age is needed. A simple correction suffices:

$$200\text{-day weight} = \frac{(\text{weaning weight} - \text{birth weight} \times 200)}{\text{age (days) at weaning}} + \text{birth weight}$$

The corrected 200-day weight should be further corrected for dam age, usually correcting to a 5-year-old dam standard. Sex corrections can also be applied or sexes assessed separately. Note that although at birth only two sexes exist, at weaning there may be bulls, steers and heifers, with weights declining from left to right in this list.

Weight of calf weaned per cow mated (WCW) is valuable for breeders selling at that age since it includes measures of production and cow fertility and it is both easy to measure and has moderate heritability. Post-weaning performance can be assessed in any of the sexes but would be particularly important in bulls. In the USA testing tended to be based on feedlot performance assessing average daily gain (ADG) over a period of 148 or 196 days. In Britain assessment tended to be based on weight at 400 days of age. In Cuba the author was involved with testing based on early weaning (90 days of age) and subsequent performance on ADG to a fixed weight of 400 kg.

Correlations between weights at different ages tend to be high and positive and the closer the two ages the higher the values. Thus, increasing size at any age is associated with increased size at other ages. Similarly, growth rate will be positively correlated to weight at given ages. A 400-day weight has been a favoured point because it is close to puberty. Increasingly the application of BLUP techniques has meant that single trait selection has been refined, but also index approaches to multi-trait selection have occurred.

Selection on carcass traits is usually directed towards reduced fatness/increased leanness either by taking ultrasonic fat measurements (e.g. at the 10th or 13th rib) or by looking at eye muscle (longissimus dorsi) area which is a highly heritable trait. It is extremely difficult to alter muscle distribution because, within a species, there are only minimal differences between breeds for this feature.

In Europe carcasses are graded on both fatness and conformation. Fat classes 1, 2, 3, 4L, 4H, 5L and 5H are in ascending order of fatness and the target area would be in the 2 to 4L range. Conformation is ranked E, U+, –U, R, O+, –O, P+ and –P in descending order of merit. Again the target area would be E to –U or possibly R. The carcass would be described with conformation first, e.g. R4L.

In Britain estimated breeding values (EBVs) are calculated for birthweight, 200-day milk, 200- and 400-day weight, backfat depth, muscling score and an overall beef value. These are done for breeds separately. The results for Charolais born in 1994 are shown in Table 9.2. In most breeds in Britain over the past two decades there has been an increase in 400-day weight as well as an increase in wither height, but also some increase in the level of subcutaneous backfat.

Sheep breeding

In most countries sheep are bred as meat producers, but all sheep breeders derive some income from wool. In countries like Australia the primary role of most sheep is as wool producers.

Wool sheep can be divided into those producing fine quality wool (the Merino breed being the best example), general purpose wool

Table 9.2   EBV results for Charolais born in 1994

Trait	Average	Top 25%	Top 10%	Top 5%	Top 1%
Birthweight (kg) ·	1.6	0.5	–0.5	–1.2	–2.2
200-day milk (kg)	–0.7	1.0	2.6	3.5	5.2
200-day weight (kg)	12.4	18.3	23.6	26.8	32.7
400-day weight (kg)	22.3	32.8	42.4	48.1	58.7
Backfat depth (mm)	0.06	–0.02	–0.09	–0.13	–0.21
Muscling score (points)	–0.03	0.11	0.23	0.30	0.44
Beef value	CH20	CH25	CH29	CH32	CH36

Source: MLC (1996)

(many British breeds fit this category) and carpet wool breeds (the New Zealand Drysdale is a specialist breed in this area, but many British hill breeds produce carpet wool quality).

Except in environmentally disadvantaged areas, most breeders would be seeking to select for some measure of reproduction such as number of lambs weaned (NLW) and possibly a measure of growth such as lamb weaning weight (LWW). Fine wool breeds such as the Merino would be selected for wool traits such as greasy fleece weight (GFW) and staple length (SL) or fibre diameter (FD). A fibre diameter of 34 microns or finer is desirable and selection would be directed towards increased fleece weight but with some upper limit on fibre diameter. Most wool traits are fairly highly heritable so that improvement of wool breeds is easier than meat breeds.

Carpet wools are heavily medullated (i.e. they have hollow fibres of hair rather than the solid fibres of wool). Breeders of such sheep would also look for weight of wool coupled with carpet wool features. Medullation is rather subjectively assessed and manufacturers do not stipulate exactly what they require. Carpet wools can be heavily, lightly or variably medullated, but more precise specifications are needed by breeders. In the Drysdale, extreme hairiness is caused by an incompletely dominant autosomal gene (N^d) which makes for easier selection.

In Britain, the sheep industry suffers not only from a plethora of breeds, many of which are of localised interest and numerically small, but also from stratification. Most sheep are hill breeds (Blackface, Cheviot, Swaledale, Welsh Mountain, etc.) which are kept in adverse environmental conditions and which must be hardy. They are usually bred pure and are not very prolific, largely owing to the environmental conditions rather than any intrinsic deficiencies. Twinning might occur for some 20% of the flock depending upon the farm conditions and in-bye land (improved land close to the buildings). Wool provides a larger portion of income in these breeds than in those discussed below, largely because lambing rates are relatively low. Flocks have the advantage of being large in many cases (over 1000).

When hill ewes begin to lose their teeth or have had perhaps five lamb crops they may be drafted out (sold off) to upland areas. Here the breeds may be what are termed longwool breeds (e.g. Border Leicester, Wensleydale, Teeswater, Hexham Blue-Faced Leicester). Longwool breeds tend to be kept in relatively small flocks (often much less than 50) and they are larger sheep than hill breeds and highly prolific, quite capable of 200 lamb crops (i.e. 200 lambs weaned per 100 ewes mated).

The usual policy is to mate draft hill ewes to Longwools, giving rise to F1 animals that are larger than hill breeds and highly prolific. The most common cross is the Border Leicester × Blackface, known as a Mule, but there are many other types. These F1 animals are mated to terminal sire breeds which cover Down breeds as well as European imported breeds such as the Charolais or Texel. In Britain the most common breed used as a terminal sire is the Suffolk.

Because the hill, upland and lowland breeders have different sheep, different conditions and different objectives the stratification system does not aid selection. The usual slaughter lamb may be 50% Down, 25% Longwool and 25% Hill, but the component parts of this have not been selected for the same traits. Hardiness is a valuable trait in the hills but difficult to assess. Because sheep tend to be crossbred there is exploitation of heterosis, but a Mule owner, for example, has minimal control over any breeding programme since his ewes are mated to terminal sires and the next generation of ewes has to be bought in from outside. Mule purchasers are dependent upon Hill and Longwool breeders for their product. Thus, a Mule user makes breeding decisions largely on where to buy his replacements and what sires to mate them to but has no control over the actual production of his Mules.

Some progress has been made in terminal sire breeds and, for example, the Suffolk breed has a sire referencing scheme in which lambs are selected for a variety of traits and on index. The component traits are 56-day weight, 140-day weight, muscle depth at 140 days and fat depth at 140 days, with all four combined into an index. A base year (1990) is used and the progress in index has been given previously (Table 5.8). Table 9.3 shows the mean values by year of birth for the four component traits.

Table 9.3 EBV values for Suffolk lambs by year of birth

Year of birth	56-day weight (kg) Eight EBV	140-day weight (kg) Scan EBV	Muscle depth (mm) Muscle EBV	Fat depth (mm) Fat EBV
1990[a]	0.005	−0.21	−0.02	−0.01
1991	0.213	0.264	0.121	0.01
1992	0.433	0.579	0.347	−0.009
1993	0.624	1.136	0.506	0.035
1994	0.794	1.565	0.581	0.037
1995	1.136	2.392	0.841	0.082
1996	1.409	2.853	1.048	0.111

[a] Base year
After Hiam (1997)

Some attention has been paid to increasing prolificacy. There is a view that the difficulty does not lie with the ability to produce high lamb crops but with the management skills needed to keep them alive. Certainly, in Britain, crops of 200 are feasible, but mean values out of F1 crosses by terminal sires would tend to be in the 170 to 190 range. At one stage Finnish Landrace sheep were brought to Britain because they were highly prolific sheep capable of producing litters of three and more. Crossing with British breeds did give rise to large litters, but lamb size was not consistent with the meat production aims of British producers. The Finn has survived largely in the Finn–Dorset which is a combination of Finnish Landrace and Dorset Horn but with only about 25% Finn.

A more interesting route lies in the discovery, in Australia, of a fecundity gene (F) in a strain of Merino called the Booroola. This strain was developed in New South Wales by the Sears brothers from the 1940s and advanced by the Commonwealth Scientific and Industrial Research Organisation (CSIRO) from about 1959. The strain was highly prolific and it was due to a single major gene. The FF animals have an ovulation rate about three times that of normal (++) ewes with heterozygotes (F+) intermediate between the two. This advantage of the FF over the F+ was not apparent at lambing, but both the FF and F+ animals excelled over the ++ ewes to the extent of something approaching 0.5 lambs. Meat sheep producers would not be interested in the Merino as such but would find the Booroola gene desirable if it could be transferred into meat breeds. In lambing terms the effect of the Booroola gene will depend upon which breed it is introduced to and the level of management.

Scrapie, a spongiform encephalopathy, is a major disease of sheep and has a genetic basis but is seemingly not inherited in the same fashion in each breed. Most breeders would wish to reduce the incidence of the disease or, preferably, eradicate it completely. The mode of inheritance in Suffolks has been discussed previously (see Chapter 9, 'General principles') and is relatively simple to control in that breed. As other breeds are studied and the genetics understood, the disease in them, too, can be controlled.

Pig breeding

Pigs as litter producers have advantages over ruminants, and the species has been advanced in genetic terms second only to poultry. Because pigs use cereals they are direct competitors to man and feed efficiency is a vital trait in view of the high proportion of total costs that is due to feeds. Over the past 30 years or more the pig has advanced in respect of reduced backfat and consequent efficiency to a considerable degree.

Backfat and eye muscle area are quite heritable traits and growth rate is related to feed efficiency. Pigs tend to be slaughtered mainly at bacon weight (90 kg) but some at pork weights (ca. 60 to 70 kg) and also at heavy hog weights (110 kg). At one time selection criteria for pork or bacon pigs were different, but now these aspects tend to be combined and slaughter weight is the principal distinguishing feature.

In Britain, pig breeding started with progeny testing, but this was quickly discontinued as performance testing was more appropriate. The MLC ran testing stations and created a triangular structure in which nucleus herds, selected on merit, formed the basis of testing.

These nucleus herds had to pass strict health tests, to conform to specific levels of performance and had to submit a number of pigs to test stations each year. Each litter to be tested had to comprise two boars, one castrate and one gilt. The gilt and castrate were killed for carcass data and the two boars were fed separately and had various measurements of performance taken. An index based on feed efficiency, growth and carcass traits was used, with each year's average being set at 100 points. Boars scoring less than 90 points were culled while those above this could be bought back by breeders. Boars over 120 were usually repurchased and most boars over 160 points entered AI stations. Because 100 was the mean each year, comparisons between years were not meaningful. A score of 100 in 1996 would be a better animal than one scoring 100 in 1990.

The increasing costs of test stations gradually undermined this programme which was not helped by breeders' reluctance (on health grounds) to take back boars that had been off their premises. The increased influence of pig breeding companies meant that nucleus herds were diminishing in both number and importance, though some actually became breeding companies. Companies did continue to test along similar lines to the MLC, but on-farm testing gradually replaced test stations.

Physical defects, particularly leg weaknesses, are crucial in pigs and potential breeding stock would be culled if defective in these features. It may be that there are more inherited anomalies in the pig than other farm species. Such defects as cryptorchidism (rare in cattle) are common in pigs along with anal atresia.

Pig companies continue to select for lean meat growth and reduced back fat, while the concept of breed is gradually disappearing into a mixture of hybrid lines or sire lines. Most testing is to a fixed weight and increasingly on *ad-libitum* diets, as opposed to restricted systems. White pigs (Large White and Landrace) are favoured, but some American breeds, developed for pork, are coloured as are some European breeds. However, white versions of the Hampshire (normally a black pig with a white band around its body) do exist. The white gene(s) can be introduced to any breed.

Group housing is favoured because pigs are social animals and derive some benefits from being in social groups. Moreover, there is evidence that boars tested in individual pens may, in about 25% of cases, develop libido problems later in life. Group feeding makes feed intake measurements harder unless electronic feeding methods are employed, without which ADG and lean production would have to serve as guides to efficiency.

Table 9.4 shows the incidence of the halothane gene in pig populations some years ago. Halothane is an anaesthetic which, when applied to pigs at 8 to 12 weeks of age, causes some of them to exhibit rigidity of the limbs. This can lead to death if the anaesthesia is not removed. It was discovered that the gene causing this (termed the halothane gene) was an autosomal recessive and that its incidence in breeds was very different across breeds (see Table 9.4).

Table 9.4 Summary of halothane positive incidence

Breed[a]	Pigs tested	Halothane positive (%)
Duroc	248	0
British Large White	764	0
American Yorkshire	225	0
American Hampshire	232	2
Dutch Yorkshire	1394	3
Norwegian Landrace	576	5
Swiss Large White	1130	6
Danish Landrace	1990	7
German Landrace (GDR)	300	10
British Landrace	1538	11
Swiss Landrace	7480	13
Swedish Landrace	1668	15
Dutch Landrace	4073	22
French Pietrain	335	31
German Landrace (GFR)	1251	68
Belgian Landrace	1260	86
German Pietrain	266	87

[a] The review paper (Webb, 1981) has more breeds given, but only those with over 220 cases are reported here
GDR, German Democratic Republic; GFR, German Federal Republic

The halothane gene is associated with beneficial effects upon lean meat yield and in ham shape which probably accounted for its high incidence in pig breeds of continental Europe where premiums had existed for well-developed hams. On the debit side, the gene was also associated with stress susceptibility, pale soft exudative (PSE) muscle, reduced litter size and sudden death syndrome. On the whole, disadvantages outweigh the benefits and when the location of the gene (in DNA terms) was discovered it meant that hh and Hh animals could be discarded from breeding lines. The percentages given in Table 9.4 will now have changed and most breeding companies seek to exclude the halothane gene from most of their lines.

Selection for reduced backfat may have gone far enough and can lead to thin sow syndrome, where females do not have enough fat cover during the lactational/breeding stage. It can also bring about reduced fat over the *longissimus dorsi*, only for wedges of fat to appear further down the ribcage and into the belly. Finally, there can be tissue separation after curing wherein the fat breaks away from the lean. Holding backfat levels at current values may be desirable while allowing greater selection for litter size.

Selection for litter size or for piglets per sow per year is increasingly important. Litter size is not highly heritable and direct selection may only lead to a genetic gain of about 0.12 piglets per sow per litter per year. The French system of hyperprolific pigs uses a policy of selecting outstanding litter-producing sows from multiplier units and placing these in a nucleus unit. This is said to lead to an advance of about 0.3 piglets per sow per litter per year. An index system developed in Edinburgh based upon assessing gilt potential from the performance of mother, grandmother and paternal and maternal half-siblings can be used on both sexes and is estimated to produce an advance approaching 0.5 piglets per sow per litter per year.

Eating quality is also receiving more attention. It has been argued that intramuscular fat is a principal component of eating quality and this has led to increased interest in the Duroc breed, which has higher intramuscular fat levels than most breeds. This breed certainly excels when used for fresh pork where intramuscular fat levels in excess of 2% seem desirable. However, for processing meat the breed does less well because levels of intramuscular fat over 3% are undesirable for this process. The different requirements of pork and the processing/ bacon market could lead to different lines being bred for different purposes.

The welfare lobby is also active in the pig industry, making it increasingly desirable to produce pigs kept in more extensive conditions. This means that group housing will be used more often and with this the opportunity to use feeding hoppers which will record the intake of each pig. Of course changing from individual to group housing may simply replace one type of stress with another and group housing can lead to 'pecking orders' and fighting. Outdoor pig production is still not a major form of production in Europe, but its increasing incidence could lead to the development of different strains/breeds from those used indoors.

The use of Chinese breeds like the Meishan is relatively new. This breed has an ovulation rate in the region of 18 as opposed to the more usual 13 and litter sizes around 13 as opposed to 10. Chinese sows have larger numbers of teats and earlier sexual maturity. Unfortunately the Meishan is vastly different in appearance from Western pigs, as well as being slow growing and having high carcass fat levels. Incorporation of Chinese breeds into maternal lines so as to improve prolificacy is under way.

Horse breeding

Studies on the racing performance of Thoroughbreds were undertaken some years ago (More O'Ferrall and Cunningham, 1974) and showed that Timeform ratings were moderately heritable (ca. 35%). Most horse races involve horses being handicapped as to the weight they should carry and this 'handicap' is such that, if perfectly arrived at, it would result in horses performing equally. These ratings are expressed in pounds of weight and would be what a horse would be entitled to receive in an average Free Handicap. Timeform ratings are adjusted annually and published for all horses at the end of the season. Other studies on racing performance put the heritabilities between 10% and as high as 74% (see More O'Ferrall and Cunningham, 1974) but with a preferred value around the mid-thirties.

A later review (Tolley et al., 1985), based on Thoroughbreds and Trotters, covered assessment on time, rankings or placings and total earnings. In general the poor data structure and sensitivity to sampling errors were important, although most authors seem to feel that mass selection could work. One problem with Thoroughbreds is the failure to allow AI to be used, though it is practised in some performance breeds (used for show jumping, dressage and cross country eventing).

Tavernier (1988) has examined the use of BLUP techniques in performance horses and has agreed with the view that while there was little advance among Thoroughbreds there was a clear genetic advance in the Selle Français and Anglo Arab breeds in France – the typical saddle horses of that country. In Sweden Philipsson *et al.* (1990) have suggested a two tier selection system for performance horses involving selection on conformation followed by performance testing.

One problem with horse breeding is the relatively small size of units, the high cost and the lack of incentive for common endeavour. If dairy cattle breeders can advance the efficiency of production they will all benefit. If racing horses became faster runners there is no similar benefit to all breeders. Races would be run quicker, but that would only benefit those studs with the fastest horses which is exactly the situation at present.

A high percentage of performance horses may indulge in show jumping (ca. 75%) but at very different levels for very different rewards. Horses which appear to excel at jumping will receive training to improve while those that do not excel are likely to remain unused for breeding. This may lead to progress, but the type of data may not lend themselves to genetic studies.

Dog breeding

The dog is both the oldest domesticated species and the most variable. Originally developed for specific purposes, many breeds no longer have access to the type of work for which they were initially bred. In general, dog breeds are divided into hounds (which hunt by smell or by sight), guard dogs, gun dogs (which essentially set up and retrieve game, particularly birds), terriers (used to kill vermin or to go down a hole to flush out or kill species like the fox), herding breeds, livestock protection breeds and, finally, utility and toy breeds which have no specific functional purpose other than as companion animals.

The difference between herding breeds, like the Border Collie, and livestock protection breeds, such as the Maremma or the Anatolian Shepherd dog, is that the former act to herd and move sheep in specific directions while the latter live among sheep and protect them from predators. Both types probably go back to the origins of sheep domestication but have developed very differently. Herding breeds play different puppy games from protection breeds and would not be suitable for the latter occupation while protection breeds with the innate instinct still require to be reared among sheep from an early age. Livestock protection breeds are currently particularly effective in the USA against coyote and to a lesser degree puma and bear. It is argued that in some locations sheep farming could not economically exist without livestock protection dogs.

Some breeds were developed as fighting breeds, either against bulls, bears or other dogs, but there is no longer room for such animals in a modern society and in most countries dog fighting is illegal. In contrast, sled dogs such as the Siberian Husky and Alaskan Malamute not only still function in their original work but also are used for sled competitions in various countries and are selected to that end, thus preserving the work potential.

Gun dog breeds tend to be relatively recent innovations (going back 100 to 150 years) and they have often developed into two distinct forms within a breed. This is particularly exemplified in Labrador retrievers. The working Labrador is relatively small, lithe and agile and very fast, while the show dog is larger, heavier and may exhibit minimal interest in, or aptitude for, work. In contrast the flat-coated retriever shows little difference between working and show strains. This division in many gun dog breeds is a disadvantage, but it is probably true that the vast majority of gun dogs no longer see a gun, still less work to one.

The advent of dog shows in the early part of the nineteenth century has increasingly led to the breeding of dog breeds for exhibition purposes based upon conformation. Unfortunately the understanding of the genetics of conformation is minimal. One South African study (Verryn and Geerthsen, 1987) on the German Shepherd dog (GSD) did examine the heritabilities of conformational measurements and found heritabilities in the medium to high range, but whether this applies to other breeds is conjecture. Dog breeders, despite working with relatively small breeding units, have succeeded in changing the shape of breeds, not always in an undesirable way though sometimes adversely.

It would be true to say that the vast majority of dogs spend their lives as companion animals, which suggests that character (temperament) is a major feature to consider. Canine behaviour is not very well studied in genetic terms, but fearfulness, an undesirable trait in most dogs, may be as high as 50% heritable. Over the years, as the role of dogs has changed, some have developed new roles. Breeds such as the GSD, Labrador and Golden retriever have found new roles as guide dogs.

Organisations such as the Guide Dogs for the Blind Association (GDBA) run their own breeding programmes with considerable success. Over the past 25 years or more that organisation has gone from success rates, in terms of dogs passing out as functional guide dogs, of around 40% to over 90%, largely because the GDBA has concentrated on what they actually wanted in their dogs rather than merely taking 'show-ring cast-offs'. In similar fashion Metpol, the dog breeding arm of the Metropolitan Police, has had considerable success in breeding GSDs as functional police animals and with improved hip status. Compared with 10 years ago more Metpol animals are found in the better hip grades.

Inherited defects probably play a greater role in dog breeding than in other livestock. There are several important eye defects (largely Mendelian) such as progressive retinal atrophy (PRA) and orthopaedic problems (mostly polygenic) such as hip dysplasia which are relatively common in some breeds. Schemes to screen for such defects do exist in many countries and, increasingly, breeders pay attention to this in breeding programmes so that culling on these counts acts as a first stage selection, albeit hampered by the fact that some defects are relatively late in onset. Many breeders of susceptible breeds will tend to screen *all* breeding stock for specific defects if schemes exist. In some areas/countries, such as Scandinavia, passing such schemes is compulsory for animals to be used for breeding, but in Britain the Kennel Club adopts an excessively tolerant view and thus hampers progress.

Canine gene mapping is going on and, increasingly, genes will be identified and located. For example, PRA in the Irish setter can now be identified by DNA evaluation, though, as yet, this same defect is not identifiable in other breeds by the same route. Some defects are associated with large size or rapid growth (elbow dysplasia), others with small breeds (patellar luxation and Legg–Calvé–Perthes disease).

The original function of the GSD (as a herding breed) has been largely superseded as the dog has become the world's principal police/army/guard dog breed. By the same token Border Collies have begun to be used in obedience and agility competitions which may have minimal bearing on their original herding function. The International Sheep Dog Society (ISDS) still controls the breeding of working sheepdogs, but those registered with the Kennel Club have no real requirements to be herders.

Failure to train and work functional dogs can have damaging consequences on behaviour. The sight hound hunting instincts of the Greyhound have been directed towards racing, but the future role of the Foxhound is in doubt if fox hunting is outlawed, although versions of this breed have been successful as Trail hounds following a man-laid track. Some breeds can make the transition to companion animal status very effectively, but this may be less true of some hounds.

Dogs still aid man in a variety of roles and specific breeds are more suited to some tasks than other breeds. There is, nevertheless, a need for breeders to decide what it is that they want the breed to do and to select for this objective. Selection for physical beauty will be of minimal value unless associated with functional ability and those breeds which have preserved working qualities should be selected to enhance, not lose, these virtues.

Although it is effective in the dog, AI is still frowned upon in some countries or permitted in such a limited fashion (e.g. Britain) as to largely defeat the objective. About 10% of males and about 30% of females are used for breeding in many breeds, but it is by no means certain that these are the upper percentiles. Application of modern breeding techniques is rarely seen, though some index selection has been practised against such traits as hip dysplasia. Too much attention is still paid to showring success and often this is achieved under judges using very conservative standards.

In some breeds, notably the GSD and other German breeds, breed surveys are undertaken in which dogs are measured and assessed against the breed standard but including some evaluation of character/work. They are placed into categories as being Recommended for Breeding or Suitable for Breeding or rejected. To the extent that these systems pick out the better animals they could help to enhance the breed. They are widely used in Germany and Australia in the GSD and have begun in Britain, albeit on a smaller scale.

Cat breeding

Cat breeding also suffers from the small breeding unit syndrome and usually operates with smaller breed populations than dogs. Cats are numerous in some countries (over 12 million in Austria and 7 million in Britain), but the pedigree animals used in organised (as opposed to random) breeding represent a small proportion of the total. The cat is very clearly a companion animal and most breeding is directed towards physical appearance and 'desirable' temperament.

As with the dog the genetics of type is unclear and most breeding is undertaken using assortative, compensatory or independent culling level methods. Great stress is laid upon head shape, coat type and eye colour and much less on actual function, with animals being judged largely in a static state at shows. Defects attract attention, as they do in the dog, but there appear to be fewer problems (or perhaps less is documented about them). Coat colour, as in the dog and horse, is controlled by a large number of genes, most with several alleles, and many breeders are conversant with the combinations or can gain access to understanding from breed literature.

In recent years some breeds have been bred towards different shapes. Oriental breeds such as the Siamese have been selected to have narrower, leggier frames and increasingly more pointed heads. In contrast, extreme forms of Persian exist in which the brachy-cephalic head structure is extreme, leading to respiratory and eye problems. Breeders appear to be free to lead their breed down such dangerous routes with minimal control, although the GCCF has started to seek to control some Persian excesses.

Dog and cat breeding in the future

There is a need in the cat and dog for breed societies to begin collecting data on a variety of features. This should not only include reproductive performance, longevity and the reasons for death but also measure the incidence of specific defects so that mode of inheritance can be established and the lines responsible noted. In view of the companion animal nature of both species, testing for character traits needs to be incorporated into breeding programmes.

Chapter 10
Breeding Policies in Developing Regions

Introduction

Animal breeding technologies in the developed world are widely accepted and fairly sophisticated, but this is less true of developing countries. Such regions are frequently characterised by having tropical or sub-tropical environments, though with considerable variation. Farming tends to be either very fragmented at a subsistence level or very extensive. Agriculture is usually important, but the sociological and political situation may be quite different from that in Europe or North America.

The major problems in such regions are usually managerial and/or nutritional. Developing nations in Latin America, Asia and Africa have about 80% of the world's human population, some 60% of the cattle, 30% of the sheep and most of the buffalo. Despite the high proportion of cattle, such regions produce only a relatively small part of the milk and meat produced in the world. Increasing the efficiency of animal agriculture, especially that of cattle and other ruminants (which are in less direct competition with man than are pigs and poultry), would clearly be beneficial to these regions. There is, however, difficulty in transferring sophisticated breeding programmes devised for the developed world to developing regions which may not possess the necessary infrastructure. Highly commercial and highly specialised systems may not be biologically, economically or even socially desirable for the developing world.

Dairy cattle

In developing nations, which are frequently tropical and range over wide ecological zones, cattle breeds have been developed which are usually of *Bos indicus* rather than *Bos taurus* origin. Such cattle are characterised by different morphological features such as long drooping ears, pronounced dewlaps and humps over the withers, but they are usually better suited to the climatic conditions. They have different sweat glands and are often more tolerant of enzootic disease. However, some of their 'adaptation' may stem from a lower productive performance.

There is some evidence that tropical breeds show much greater variation in milk yield than European breeds and that maternal instincts towards the calf are more pronounced with consequent difficulties with milk let-down if the calf is not present.

Selection studies undertaken among tropical cattle have tended to be based upon small numbers or undertaken over a short time period. Genetic variation certainly exists among such cattle for many economically important traits, but selection within native breeds, which are late maturing and low milk producers, is likely to be unrewarding.

On the other hand, replacement with temperate livestock can be fraught with dangers. Imported European/North American breeds can be successfully farmed in the tropics when there is substantial control of disease and feeding and management standards are high, with climatic stress reduced to a minimum. Such conditions are likely to prevail only in special circumstances and, in the main, temperate imports can exhibit greater susceptibility to disease, reduced growth, poor fertility and high mortality (see review by Pearson de Vaccaro, 1990). Replacement techniques are likely to succeed only in special circumstances when linked with improved managerial skills. These will, of course, apply in some locations.

Replacement can be undertaken by the use of semen and AI which means that not only the initial introduction is relatively cheap but also the first produce are crosses between native cows and semen of temperate-type bulls. The difficulties with AI are that in many developing regions the techniques may not be as advanced as they are in advanced countries and there may be inadequate levels of training for AI personnel. The importation of semen usually requires hard currency which may be another stumbling block. In addition, roads and communications may be poor and there may be defective nutrition as well as disease among cattle. These problems will apply to an even greater degree if one is contemplating the importation of frozen embryos.

In a review of the subject, Cunningham and Syrstad (1987) showed that the introduction of up to 50% *Bos taurus* genes into *Bos indicus* cattle tended to improve performance in most traits, but beyond this point the calving interval increased due to reduced fertility (see Figure 10.1).

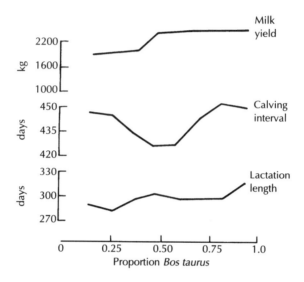

Fig. 10.1 Performance of different *Bos taurus/Bos indicus* combinations (after Cunningham and Syrstad, 1987)

Any crossbreeding programme needs correct evaluation which means that the overall picture must be examined rather than simply one feature. Crossing Sahiwal with Holsteins may give an increased milk yield even if taken beyond the 50% Holstein level, but continued upgrading might bring longer calving intervals, more reproductive disorders, increased disease susceptibility, as well as disadvantages in fringe areas like meat production or draught capacity. Pearson de Vaccaro's (1990) study of survival rates showed imported breeds to average 2.6 calvings versus 3.1 for local breeds. However, better results were seen from crossbreds containing 50 to 62% imported 'blood'.

The specialist dairy cow of Europe or North America is exactly that – *specialist* – but in the tropics a dairy cow may be used not only to produce milk but also beef through her calf, draught and as a source of fertiliser through her dung. In some areas productivity is not easy to assess especially if, as in parts of Africa, herds are nomadic.

In some regions of Latin America it is policy to milk the cow once or twice daily and then allow the calf access to the cow after each milking or on a constant basis. The cow thus provides milk as well as rearing her own calf to about 6 months of age. These systems usually result in better calf survival than with artificial rearing and may increase total milk yield. However, milk recording is clearly difficult to evaluate and breeding objectives are obviously different from those in temperate regions.

There is some evidence that heterosis between native and exotic cattle is greater in poor than in good environments and that heterosis is likely to be higher between *Bos taurus/Bos indicus* crosses than between two *Bos taurus* types. Most experts would agree that some influx of exotic 'blood' is needed, but without any hard and fast rule as to the degree of inclusion. AI is likely to be the best way of incorporating exotic breeds but is not necessarily easy for reasons already given above. Small herd size makes organised schemes difficult to instigate and there is the need for government aid in recording and perhaps the setting up of some kind of nucleus unit.

In good environments – and the tropics are extremely variable even within a single country – there may be a good reason for using *Bos taurus* cattle either by importation or by grading-up. In other situations the choice may be restricted to a degree of exotic 'blood' incorporated either by rotational crossbreeding or by the creation of a new breed as was done to form the Siboney de Cuba (see Figure 8.5). With the dual purpose system of milk plus a suckled calf the breed type will probably be a crossbred rather than an exotic. Objectives may be readily defined, e.g. 1800 kg milk plus a 180-kg calf plus a fertility level of 85%, but selection for such a composite is not easy.

Beef cattle

Beef cattle in tropical areas are a somewhat easier proposition than dairy cattle. Again in these regions they tend to be of zebu (*Bos indicus*) type. However, in parts of Latin America breeds exist that are of *Bos taurus* origin but which have been there since the time of the *conquistadores*. Under the generic title of Criollo these include such breeds as the Blanco Orejinegro (BON), the Romosinuoano and the Costeño con Cuernos. Although European in origin these breeds are virtually indigenous by long association with the region. Criollo cattle may be more fertile than zebu types which are unlikely to present oestrus much before 22 to 24 months while temperate cattle will do so around 15 to 16 months, even in tropical conditions. In Mexico the term Criollo is often used to indicate 'mixed' blood cattle rather than the breed types mentioned above.

Beef cattle production levels tend to be poor with calf crops around the 40 to 60% level compared with 85 to 95% in Europe. Seasonality of herbage does not aid the situation. Calving intervals are closer to 460 days rather than the 380 days of Europe albeit with a wide range around this figure. Age at first calving is likely to be closer to 3 years in imported *Bos taurus* cattle and closer to 4 years in native types. Individual herds can do very much better than the average in all these traits but such units are few and far between, and although exotic breeds can drastically reduce age at first calving as well as calving interval, the reduction in the latter is largely only true of the first interval and less so of subsequent ones.

Dystocia is rarely a problem in native breeds where calves are often quite light at birth (below 30 kg), but it is well established that mortality tends to be higher in small calves than in the more standard sized calf of 36 kg. Crossing with European breeds will increase birth weight and could increase dystocia to a degree. Tropical breeds are notoriously difficult to get in calf while lactating as opposed to when dry which does not help reproduction. Mortality is high in most tropical cattle (minimal values of 10% to weaning at 7 or 8 months), but temperate breeds may not necessarily fare better.

In many situations there is a need to examine the potential of native breeds before seeking to import new ones. In South and East Africa the Africander was widely used in preference to local breeds, but under scientific scrutiny indigenous breeds like the Tuli and Mashona have performed better (see Hetzel, 1988). The Brahman has been advocated to improve African cattle but justification from such trials as exist is hard to find.

Beef cattle breeding in tropical situations can be done by selection within local breeds using similar traits to those in Europe, but because of reproductive rates the need to look at *kilograms of calf weaned per cow mated* is crucial. There is evidence that temperate breeds as well as Brahmans can be effective as crossing sires and crosses with such native breeds as the Tuli and Mashona need to be studied. Such evaluations may be best undertaken in the local conditions where the cattle will be farmed.

In Latin America Plasse (1983) showed advantages from crossing exotic breeds with native zebu cattle but less advantage from crossing Criollo breeds with zebu types beyond the F1 generation. There may be evidence to suggest better results from European breeds as opposed to British ones but more data are needed. The need to preserve some of the Criollo breeds is important and crossing them out of existence may be a short-sighted and erroneous policy.

Pigs

The pig lends itself better to tropical conditions than most farm species and there are numerous pig breeds developed in tropical regions. Most of the criteria thought desirable in temperate areas would also equally apply to the tropics. One would thus be seeking to increase litter size, frequency of farrowing, survival ability and lean tissue growth. Some of the Chinese pig breeds are extraordinarily prolific, but, in common with many of the breeds of Central and South America, they tend to be lard (fat) breeds with undesirable carcass traits.

It is probable that most breeds will express themselves best in the environment in which they were selected, especially if, as seems likely, genotype–environment interactions are more likely in the pig than in cattle. Imported mature stock are likely to suffer in the tropics and be less fertile – at least for some time. Boars may show a reduced libido and females may either produce smaller litters or show increased mortality, especially if, as seems likely, milk yield is depressed. On the other hand, carcass improvements would be better achieved by the use of imported developed breeds and grading-up would seem a sensible policy.

Disease will be a major drawback in some tropical areas and the ability to produce disease-free strains as in temperate regions would be hampered by the cost factor.

Sheep

Sheep are not ideally suited to tropical regions partly because wool is to some degree disadvantageous in such areas, and also because of photoperiodicity. Imported sheep have some difficulties with reproduction when day length is fairly static. As a consequence the importation of temperate breeds is rarely practised unless into high altitude areas.

Adapted sheep can cope well with the lack of major day length changes and some tropical breeds (e.g. the Barbados Black Belly) are highly prolific. Many tropical breeds are hair sheep rather than wool producers and most tend to be small which is contra-indicated with meat production in temperate regions but may be less crucial in developing areas. Wool/hair is not usually a primary product but will be a useful by-product in Africa and Asia while milk sheep are important in some areas, notably the Middle East. In developed countries selection for increased lean is a vital part of meat production, but in tropical areas carcass fatness may be less of a disadvantage. In many locations flocks would be small which does not lend itself to genetic improvement.

Chapter 11
Breeding Schemes and Future Developments

Introduction

Few animal breeders operate on a very large scale. The average size of dairy herds varies throughout the world but in most locations would be less than 50 cows. Pedigree beef herds in Europe might be closer to 10 or 20 cows though they might be very much larger in the ranching situations of the Americas or Australia. Pig herds may, in many instances, exceed 100 sows while sheep flocks can be numbered in thousands, especially in hill situations in Britain, and can be even larger in Australia, where numbers can run into many thousands. Dog breeders often operate with no more than a handful of bitches and kennels with more than 20 are exceptional. With dog breeding there is the risk that large numbers may be incompatible with the socialisation necessary to preserve good character. Catteries are similarly small and horse studs may be relatively tiny.

Because genetics is essentially a numbers game the relatively small size of breeding units places great restriction on the progress any single breeder can expect to make. In most farm species the genetic progress expected stems, in large part, from the fact that government or semi-government bodies have become involved as well as commercial companies. Poultry breeding is almost exclusively in the hands of large commercial companies and most producers simply buy day-old chicks from the company of choice, replacing them once they have been reared to slaughter age or laid eggs for the required cycle. Increasingly, pig breeding is following this example as breeders purchase gilts/boars from commercial companies which effectively control the breeding programme. Cattle and sheep breeders have remained outside this sphere of influence, although organisations controlling AI such as dairy boards are increasingly managing the sire end of the dairy and beef market and may do the same for sheep once AI becomes more efficient and acceptable in that species.

Group breeding schemes

Increasingly, private breeders are being forced to consider cooperative ventures of one kind or another. While pedigree breeders may retain the element of commercial and breeding competition with their colleagues, they are increasingly aware that cooperative schemes, either through their breed societies or in collaboration with fellow breeders, are becoming more necessary. One result has been the development of group breeding schemes.

Both sheep and beef cattle have benefited from such schemes, often called nucleus schemes. They date back to the New Zealand Romney scheme of 1967 and have expanded since then into some very large units. Mainly applied in Australia and New Zealand there are some sheep schemes which encompass over one million ewes.

In essence group schemes involve a number of breeders who agree to cooperate using their own flocks/herds which can be pedigree, commercial or a mixture of the two. Usually, a nucleus unit is established on a separate unit or on one of the cooperating farms. This nucleus is formed from the 'best' females from each of the cooperating farms. What exactly 'best' means will depend upon the criteria chosen for selection. If only pedigree stock are used there will be a two-tier system, but with the addition of commercial animals this becomes a three-tier system.

Groups can vary enormously in both size and objectives, but the essential format involves intensive recording and measurement in the nucleus unit. The very best males produced are used in this unit and replacement females also stem, in part, from the nucleus. In addition, females are sent to the nucleus from the cooperating units annually. Cooperative units receive nucleus males. Because there is a two-way flow of stock (and genes) both into and out of the nucleus unit it is termed an open nucleus. An example of a group breeding system is shown in Figure 11.1 using hypothetical figures.

In general, nucleus units contain about 10% of the total size of the cooperating farms and take about half their replacement females from these farms each year. It is thought that such units can produce

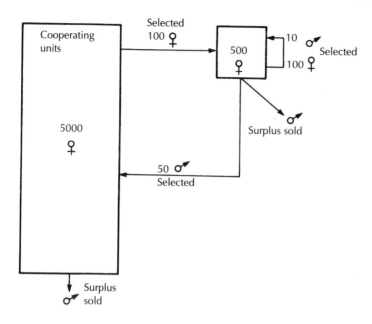

Fig. 11.1   Group breeding scheme

10 to 20% greater progress than could individual breeders working in isolation.

Such schemes have the advantage of continuity (they can exist if a member withdraws or dies) and group identity and, by virtue of larger financial size, can provide improved management, technical assistance and customer services while at the same time providing better data collection, shorter generations and more high-powered computer assistance.

On the debit side the nucleus unit is potentially subject to disease problems, the more so because of constant influx of new stock. There may be difficulties in assessing genetic merit in the different units and there is the constant danger of trying to screen incoming females for too many features which adds to cost. Closed nucleus units would avoid some of the health risks but would also make less theoretical progress.

Despite their success in Australasia such schemes have been rare in Britain. A Welsh Mountain sheep unit exists and the Suffolk breed had a kind of nucleus at Edinburgh University for some time. One of the problems in Britain is the stratified nature of the sheep industry.

Artificial insemination

By no means a new technique since it traces back to the nineteenth century and has been in wide commercial use since the 1940s, AI has been a major tool in the advancement of dairy cattle. Without AI the testing of dairy sires and the application of BLUP techniques would have been much harder. In addition, it has been AI which has allowed the changing of breeds by grading-up/migration, not only within countries but also between them. This is exemplified in Britain where in the space of 30 to 40 years the whole structure of cattle breeds has changed. The Shorthorn gave way to the Friesian which, in turn, is being superseded by the Holstein. On the beef side the influx of the Charolais in 1960 served as a stimulus to the beef breeding industry and many traditional beef breeds have become rare as they have been replaced by breeds from continental Europe.

Although technically as effective in beef as in dairy cattle the nature of beef cattle management has not permitted the application of AI as extensively as in dairy cattle. Initially, breed registration rules acted as a dampener on the use of AI but that is no longer true. Nevertheless, cheap and effective synchronisation techniques would help to further AI use. In sheep the difficulties with freezing of semen and the relative cheapness of sires have not helped to spread AI and this is to some extent true in pigs in Britain though many private breeding companies use AI routinely.

Sexing of semen will be a step forward, principally in cattle, but, as yet, the current success rate is rather illusory. The advantages of sexing in terms of enhancing choice would be great, leading to 'all-male' production in beef herds and a preponderance of females in dairy herds.

Embryo transfer

The role of MOET schemes has been explained (see Chapter 9) and the techniques have become easier and increasingly non-surgical. As success rates increase and cost comes down, greater commercial application of embryo transfer will take place and will be enhanced by the sexing of embryos. Embryo splitting can lead to increased pregnancy rates and the application of twinning in cattle without the problems of freemartinism could be a step forward.

Genetic engineering

The mapping of genomes has been mentioned earlier (see Chapter 4, 'Gene mapping') and is well advanced in several farm species as well as occurring in the dog. The identification of the location of specific genes clearly has value in respect of deleterious alleles as well as in respect of beneficial ones. The identification of PRA in the Irish setter can lead to the eradication of this major blindness gene. In the same way, the gene for blood group B21 can bring resistance to Marek's disease in poultry. Similarly gene K-88 can bring resistance to *E. coli* in pigs. In man the location of cystic fibrosis and Huntington's chorea genes will prove valuable in affected families.

Identification of genes may be the first step in genetic engineering and the simple removal of implicated animals will enhance the breed by eradicating deleterious genes. However, in some cases there will be a need to transfer genes into individuals or into other species. Transgenic animals so formed can be useful though not always is such a transfer free from side effects. Early work putting growth genes into pigs were associated with arthritic side effects, but more recent work putting human protein genes into animals for recovery in animal milk has been effective. Insulin can now be obtained from sheep as can certain blood factors useful in the treatment of haemophilia. These therapeutic uses are highly important but do not really constitute agricultural advances.

In the next decade or more, advances in the identification of major deleterious genes will increase and this will lead to their reduction in affected breeds. However, in animal breeding most of the really important traits are polygenically inherited and searching for quantitative traits will be somewhat harder. Tracking the inheritance of what are called quantitative trait loci (QTLs) will be increasingly studied and gene mapping will be enhanced and have greater practical application. At the same time we will increasingly know the number and effects of genes affecting such traits which will help selection programmes. Studies of breeds as diverse as the large White and Meishan should allow examination of major loci implicated in litter size, age at onset of puberty and fatness levels and these are well underway.

Cloning of animals has already been achieved and the creation of the famous Finn–Dorset 'Dolly' at the Roslin Institute, Edinburgh, in the mid-1990s has been widely reported. However, the publicity involved has tended to be based upon the consequences of 'allowing' this in man rather than on what could be done in animals. This, in turn, could lead to the 'ethical lobby' seeking restriction on not only what one might do in man – with which many will concur – but also what will be permitted in animals.

In 1988 Dr John King said that the idea of gene transfer may be 'a false dawn which opens up new opportunities yet to be realised' (King, 1988). At present much research is being done in this area of genetics and it is impossible to believe that none of this will bear fruit. There will be enormous benefits for some species, but whether QTL work will be as effective or as quick as some hope remains to be seen.

Preserving genetic variation

Many breeds of farm livestock have declined in importance and some have become extinct. Usually this has occurred because the populations were always small and struggled to survive in the face of inbreeding depression induced by their very smallness. Some have been popular breeds which have lost their popularity as criteria changed and other breeds were found to be commercially more viable for the needs of a particular era.

Some organisations have sought to preserve these rare breeds by farming them while others have stored semen or embryos: other breeds are preserved in zoos. The Food and Agriculture Organisation (FAO) of the United Nations has begun to catalogue such breeds.

Preservation of small groups is always difficult because their very smallness makes inbreeding inevitable while disease could wipe them out at a stroke. Moreover, if these rare breeds are less commercially viable than the breeds which replaced them, the farming of rarities is done at an economic disadvantage. Some would argue that these out-of-date breeds serve no useful purpose and should be allowed to wither and die. Others argue that demands may change and today's pariah may be tomorrow's salvation. It is often claimed that rare breeds may have some valuable feature that will one day be needed and thus preservation on this 'off chance' is essential.

It is certainly true that, in Britain, breeds like the Jacob sheep have had a new lease of life in the general climate of what is called sustainable agriculture, where there is something of a regression to more extensive farming. This may preserve some rare breeds but is unlikely to bring them into prominence. Unless a breed can be shown to have some economic merit it will have little future beyond its own rarity. Whatever one's personal feelings the sad truth is that domestic breeds, just as with wild species, will become fewer as some go to the wall. Rare breeds will simply tend to get rarer and eventually die out unless there is government and genetic help to preserve them. The likelihood is that more breeds will join the rare and at risk categories as they fail to compete with the popular breeds. A breed like the Holstein is flourishing, but its very success means harder times for other dairy breeds.

Appendix 1
Coefficient of Inbreeding

The coefficient of inbreeding was devised by Sewell Wright in the 1920s and is an attempt to measure the degree of homozygosity in an individual. The formula is:

$$F_x = \Sigma 1/2^{n_1 + n_2 + 1} \, (1 + F_A)$$

where F_x is the inbreeding coefficient, n_1 is the number of intervening generations from the sire to the common ancestor and n_2 is the number of intervening generations from the common ancestor to the dam. F_A is the inbreeding coefficient of the common ancestor.

Most textbooks use this formula and recommend arrow diagrams to construct pedigrees. This is acceptable as long as pedigrees are not complex, but with complex pedigrees arrow diagram construction can lead to many errors. It is thus better to use the technique of Willis (1968) which alters the formula to:

$$F_x = \Sigma 1/2^{n_1 + n_2 - 1} \, (1 + F_A)$$

where n_1 and n_2 represent the actual generations, counting the parents as generation 1.

Using the pedigree of Roan Gauntlet (Chapter 6) we can describe this bull as inbred:

Champion of England 3.3/2
Lord Raglan 4.4/4

This means that Champion of England appears twice in the third generation of the sire's side and once in the second on the dam's side. Similarly, Lord Raglan appears twice in the fourth generation on the sire's side and once in the fourth on the dam's.

The inbreeding to Champion of England is

$$1/2^{3+2-1} = 1/2^4$$
$$\text{plus } 1/2^{3+2-1} = 1/2^4$$

which is $0.0625 + 0.0625 = 0.125$ or 12.5%. Since there is no knowledge of the actual inbreeding of Champion of England the part $(1 + F_A)$ is ignored.

The inbreeding to Lord Raglan is

$$1/2^{4+4-1} = 1/2^7$$
$$\text{plus } 1/2^{4+4-1} = 1/2^7$$

which is $0.0078 + 0.0078 = 0.0156$ or 1.56% and again we ignore the $(1 + F_A)$ portion.

The total inbreeding is thus $0.125 + 0.0156$ or 0.1406 (14.06%).

If Champion of England had been inbred let us say 11.5% then the inbreeding to that bull of 0.125 would need to be multiplied by $1 + 0.115$ (1.115) and thus would have been 0.1394 or 13.94%. Note that in multiplication the proportion value should be used, not the percentage value. Multiplying by $1 + 11.5$ would result in a value over unity and coefficients cannot exceed 1.00 or 100% and in livestock would rarely approach this level. The common ancestor inbreeding is not usually important unless the common ancestor is inbred around 10% or more.

The advantage of the above technique is that there is no requirement to rewrite the pedigree, which is used exactly as presented. The risk of confusion in drawing complex arrow diagrams is also avoided. Care has to be taken when inbreeding is to a specific ancestor with secondary inbreeding to animals behind this ancestor. Ways of dealing with this appear in the original paper (Willis, 1968).

Appendix 2
Coefficient of Relationship

The simplest way to construct a coefficient of relationship between two animals (A and B) is to calculate the coefficient of inbreeding (as explained in Appendix 1) that would result from mating A and B and then doubling this figure. It does not matter if A and B are of the same sex since you are only calculating the theoretical inbreeding, not actually making a mating. If A and B mated together gives an inbreeding coefficient of 0.079 then the coefficient of relationship would be 0.158 or 15.8%.

Glossary

Ad libitum: unrestricted.

Additive: that part of the variance which can be transmitted to the next generation.

Allele: any of the alternative forms of a gene.

Artificial insemination (AI): the collection of male sperm and its placing in the female reproductive tract.

Artificial selection: selection made by man.

Assortative mating: mating animals which look or perform the same.

Autosomes: those chromosomes which are not sex chromosomes.

Backcross: cross between an F1 and either parent.

Biometrics: statistics as related to biological data.

Birth type: whether born a single or twin, etc.

Bloodline: usually refers to animals with some degree of relationship.

Breeding value: the merit of an animal as a breeding prospect.

Castration: removal of the testes (such animals are termed castrates).

Cell: basic unit of living organisms.

Chromatids: those parts of the chromosome after it has split longitudinally.

Chromosomes: thread-like structures found in the cell nucleus on which genes are carried.

Coefficient of variation: statistical term. Standard deviation expressed as a percentage of the mean.

Common ancestor: an individual appearing on both sides (sire and dam) of a pedigree.

Compensatory growth: growth occurring during a period of unrestriction after a period of restriction.

Contemporary(ies): animal(s) born and reared at the same time (more or less) as those being studied.

Control(s): an unselected population against which selected ones are compared.

Correction factors: amounts added to a trait to make it comparable to other measurements. For example, to correct weight in a female to what it would have been as a male. Sometimes multiplication factors are used rather than additive ones.

Correlated response: effect on one character of selection undertaken for another.

Correlation: statistical term (r) describing extent of an association between two traits.

Crossbred: an animal produced from parents of different breeds.

Crossing-over: the exchange of parts of one chromosome with its homologous partner.

Cryptorchid: failure of a testicle to descend (unilateral) or both testicles to descend (bilateral) into the scrotum.

Culling: removing poor or unsuitable animals from the population.

Deleterious gene: gene (usually simple) causing the animal to be defective in some way.

Deoxyribonucleic acid (DNA): the chemical that is the basis of the gene.

Deviation: statistical term indicating the difference between an observation and the mean of the population.

Diploid: the normal number of chromosomes for a species.

Distribution: statistical term describing the spread or range in observations.

Docking: removal of part of an animal's tail.

Dominant: refers to a gene which operates when present only in single dose, i.e. there need only be one gene for it to work.

Dressing percentage: carcass weight as a percentage of the live animal weight.

Dropsy: an inherited defect where tissues fill with fluid.

Dystocia: difficulty in giving birth.

Embryo: the early stage of an organism in the uterus.

Embryo transfer (ET): technique involving removal of an embryo from one female (the donor) and placing it into another (recipient).

Enzymes: substances which trigger off reactions but are unchanged by those reactions.

Epistasis: interaction between genes (alleles) at different loci.

Fecundity: number of progeny born and reared.

Fertility: ability to conceive (female) or produce viable sperm (male).

Fetus: unborn young still in the uterus.

Follicle: structure in the skin from which hair fibres grow.

Fitness traits: those concerned with reproduction and viability.

Frequency: statistical term indicating number of times an observation occurs.

F1: first filial generation (first cross).

F2, F3, F4, etc.: subsequent generations, e.g. F2 is F1 × F1.

Gamete: reproductive cells (ova, sperm) which unite to form the zygote.

Gene: the basic unit of inheritance.

Generation interval: average parental age when progeny are born.

Genetic drift: changes in gene frequency as a result of chance or random effects.

Genetic engineering: modification of an animal's genetic constitution by manipulation of genes.

Genetic gain: heritability/selection differential.

Genetic isolate: strain or line that is distinct from others.

Genotype: the genetic make-up of an animal.

Genotype–environment interaction: change of ranking order of different genotypes in different environmental conditions.

Germ cell: gamete.

Germ plasm: genetic material in an animal.

Haemophilia: blood disease(s) affecting clotting time.

Half-bred: first cross (F1) between two different breeds.

Haploid: half the number of chromosomes typical for the species. This number is seen in eggs or sperm.

Heritability: h^2; the additive variance as a proportion of the total variance.

Heterosis: superiority of a cross over the mid-parent.

Heterozygote: an animal with unlike alleles at a locus (e.g. Bb).

Hogg (hogget): sheep aged 6 to 12 months of age (approximately).

Homologous: of common origin. Chromosomes of a pair.

Homozygote: an animal with like alleles at a locus (e.g. BB, bb).

Hormone: secretion from special glands which permits or encourages certain functions.

Inbreeding coefficient: the rate at which homozygosity is increased.

Inbreeding depression: decline in performance due to inbreeding.

Joining: putting males with females with a view to their mating.

Karyotyping: examination of chromosomes.

Kemp: coarse, heavily medullated fibres seen in a fleece.

Killing-out percentage: see *Dressing percentage.*

Let-down: act of releasing milk from the udder, controlled by oxytocin.

Lethal gene: a gene which when expressed causes death.

Linkage: genes associated because they appear on the same chromosome.

Locus (plural *loci*): location on a chromosome for a specific gene.

Maintenance: feed required for an animal's basic functions when not performing a productive function like growth/lactation/reproduction.

Mean: total value of the observations divided by the number of observations.

Medulla: cavity or hollow inside a fibre of sheep hair.

Meiosis: reduction division from diploid to haploid state.

Migration: introduction of animals from one population (or breed) to another.

Mitosis: cell division to form two new but identical cells.

MOET: multiple ovulation embryo transfer.

Monozygous: originating from the same egg.

Mutation: change in genetic material induced naturally or by certain chemicals or by radiation.

Natural selection: selection brought about by nature.

Nicking: used when a specific cross appears to be successful.

Normal distribution: bell-shaped curve which describes variation seen in polygenic traits.

Nucleus: centre of a cell in which the chromosomes are found.

Objective trait: one that can be defined and measured with some precision.

Oestrus: heat period in a female during which mating can occur.

Ovary: female reproductive organ.

Overdominance: when the heterozygote outperforms both homozygotes.

Parity: order of birth, e.g. first calving, second calving.

Pedigree: record of the ancestors of an individual.

Performance test: assessment of the animal's own performance.

Perinatal: around birth.

Phenotype: outward expression of the animal's genetic make-up.

Placenta (afterbirth): structure in uterus which surrounds the fetus and through which it is fed. Expelled after birth.

Plateau: term used to describe state where selection ceases to produce further progress.

Pleiotropy: gene having an effect upon more than one trait.

Polygenic: controlled by many genes.

Population: group of individual animals.

Prepotency: ability of an animal to reproduce its own features.

Progeny test: evaluation of an animal on performance of its offspring.

Puberty: sexual maturity.

Qualitative traits: traits that are distinct or discrete and can be counted, rather than measured. Often simply inherited.

Quantitative traits: traits that are measured and which show continuous variation. Usually polygenically inherited.

Random: by chance.

Recessive: allele that has to be present in duplicate to be obvious. One that is masked by a dominant allele.

Reciprocal cross: cross made in both directions, e.g. Charolais male to Hereford female and vice versa.

Reduction division: see *Meiosis.*

Reference sire: sires of known merit whose progeny are interspersed in other herds or flocks to act as a standard for comparison.

Regression: statistical term (b) measuring by how much one trait changes for each unit change in another trait.

Relative Economic Value (REV): estimate of value (financial) of one trait relative to others.

Repeatability (R): extent to which a trait is repeated next time around.

S/P ratio: ratio of secondary to primary follicles in wool (measure of quality).

Segregation: separation of alleles of a pair to form germ cells.

Selection differential: difference between mean of parents and mean of population from which they were drawn.

Selection index: selection for multiple objectives by pooling information into a single score or index.

Selection intensity: severity with which selection is practised, e.g. proportion used for breeding. Mathematically, the selection differential/phenotypic standard deviation.

Semen: male sperm and lubricating fluids produced by the testes.

Sex chromosomes: that pair of chromosomes determining sex (XX and XY in animals).

Sex controlled (influenced): condition seen more frequently in one sex than the other.

Sex-limited: condition expressed only in one sex.

Sex-linked: condition carried on the sex chromosomes.

Sibling (sib): brothers and sisters. Full sibs have the same parents, half sibs have one parent in common. May not be from the same litter.

Skewed: distribution that is distorted with a long tail to one side.

Standard deviation: statistical term describing range of variation around a mean. Square root of the variance.

Stud: term used to describe a herd/flock of pedigree animals usually of some fame. In Britain used to describe a male offered for breeding.

Subjective trait: one defined in a non-precise way usually assessed by a score or on a hedonic scale.

Superovulation: stimulation of the ovary to produce more eggs than normal.

Teaser: vasectomised male used to detect females in oestrus.

Telegony: discredited belief that crossbreeding a female will affect all her future offspring from future matings.

Test-mating: mating a suspect carrier of a deleterious gene to an animal known to have that gene in order to check genetic status of the testee.

Threshold trait: one which is seen in limited forms (usually two or three) but which is polygenically controlled.

Top-cross: using a sire of the same breed but new bloodlines.

Trait: characteristic or feature of an animal.

Truncation point: performance level at which selection is made and some animals culled.

Type traits: those associated with physical features relating to a standard of excellence usually drawn up by a breed organisation.

Ultrasonics: equipment that identifies fat layer, muscle area or fetus, using high frequency sound waves.

Uterus: female organ in which fetus develops.

Variance: statistical term to measure variation in a population. Square of the standard deviation.

Yearling: animal of about 12 months of age but under 24 months.

Zygote: product of union of two gametes.

References

Becker, W. (1984) *A Manual of Quantitative Genetics*, 4th edn. Academic Enterprises, Pullman, Washington.

Cameron, N.D. (1997) *Selection Indices and Prediction of Genetic Merit in Animal Breeding*. CAB International, Wallingford.

Cunningham, E.P. & Syrstad, O. (1987) Crossbreeding *Bos indicus* and *Bos taurus* for milk production in the tropics. *FAO Animal Production and Health Paper* No. 68.

Dalton, D.C. (1985) *An Introduction to Practical Animal Breeding*, 2nd edn. BSP Professional Books, Oxford.

Falconer, D.S. & Mackay, T.F.C. (1996) *Introduction to Quantitative Genetics*, 4th edn. Longman, Harlow.

Gibson, J.P. (1987) The options and prospects for genetically altering milk composition in dairy cattle. *Animal Breeding Abstracts*, **55**, 231–43.

Henderson, C.R. (1949) Estimation of genetic changes in herd environment. *Journal of Dairy Science*, **32**, abstract 706.

Henderson, C.R. (1973) Sire evaluation and genetic trends. In: *Animal Breeding and Genetics*. American Society of Animal Science/American Dairy Science Association, Champaign, Illinois.

Hetzel, D.J.S. (1988) Comparative productivity of the Brahman and some indigenous Sanga and *Bos indicus* breeds of East and Southern Africa. *Animal Breeding Abstracts* **56**, 243–55.

Hiam, D.R. (1997) *Ram Buyers Information Sheet*. Suffolk Sire Reference Scheme, Powys.

Holstein Friesian Society (1997) Tables. The Holstein Friesian Society, Rickmansworth, Herfordshire.

King, J.W.B. (1988) The future role of the new technologies – what opportunities do they offer? *Conference on Harnessing the New Technologies for Profitable Beef Breeding and Production*, Stoneleigh, Warwickshire, 1988.

Lamberson, W.R. & Thomas, D.L. (1984) Effects of inbreeding in sheep: a review. *Animal Breeding Abstracts*, **52**, 287–97.

Lauprecht, E. (1961) Production of a population with equal frequencies of genes from three parental sources. *Journal of Animal Science*, **20**, 426–32, also errata p. 902.

MLC (1995) *Beef Yearbook*. Meat and Livestock Commission, Milton Keynes.

MLC (1996) *Beef Yearbook*. Meat and Livestock Commission, Milton Keynes.

Mohiuddin, G. (1993) Estimates of genetic and phenotypic parameters of some performance traits in beef cattle. *Animal Breeding Abstracts* **61**, 495–522.

More O'Ferrall, G.J. & Cunningham, E.P. (1974) Heritability of racing performance in Thoroughbred horses. *Livestock Production Science*, **1**, 87–97.

Newton Turner, H. & Young, S.S.Y. (1969) *Quantitative Genetics in Sheep Breeding*. Macmillan, Melbourne.

Nicholas, F.W. (1979) The genetic implications of multiple ovulation and embryo transfer in small dairy herds. *Proceedings of the Conference of the European Association for Animal Production*, Harrogate.

Nicholas, F.W. (1987) *Veterinary Genetics*. Clarendon Press, Oxford.

Nicholas, F.W. & Smith, C. (1983) Increased rates of genetic change in dairy cattle by embryo transfer and splitting. *Animal Production*, **36**, 341–53.

Pearson, K. (1931) *Tables for Statisticians and Biometricians Part II*. Biometric Laboratory, University College, London.

Pearson de Vaccaro, L. (1990) Survival of European dairy breeds and their crosses with zebus in the tropics. *Animal Breeding Abstracts* **58**, 475–94.

Philipsson, J., Arnason, Th. & Bergsten, K. (1990) Alternative strategies for performance of the Swedish warmblood horse. *Livestock Production Sciences*, **24**, 273–85.

Plasse, D. (1983) Crossbreeding results from beef cattle in the Latin American tropics. *Animal Breeding Abstracts*, **51**, 779–97.

Ponzoni, R.W. & Newman, S. (1989) Developing breeding objectives for Australian beef cattle production. *Animal Production*, **49**, 35–47.

Tavernier, A. (1988) Advantages of BLUP animal model for breeding value estimation in horses. *Livestock Production Science*, **20**, 149–60.

Tolley, E.A., Notter, D.R. & Marlowe, T.J. (1985) A review of the inheritance of racing performance in horses. *Animal Breeding Abstracts*, **53**, 163–85.

Verryn, S.D. & Geerthsen, J.M.P. (1987) Heritabilities of a population of German Shepherd dogs with a complex interrelationship structure. *Theoretical Applied Genetics*, **75**, 144–6.

Webb, A.J. (1981) The halothane story in pigs – lessons for poultry. *33rd Poultry Round Table*, Edinburgh, 1981.

Willis, M.B. (1968) A simple method for calculating Wright's coefficient of breeding. *Revista cubana Ciencia Agricola* (English edition) **2**, 171–4.

Willis, M.B. (1989) *Genetics of the Dog*. H.F. & G. Witherby, London.

Further Reading

In addition to the publications cited in the References the following are recommended to readers seeking to delve deeper into genetics per se or into a particular species. The editions listed are the latest that I have. In some cases later editions may exist and this should be checked before purchase.

General genetics books

There are numerous general genetics books and, in view of the rapidly increasing knowledge in the field, they can become quickly dated. The following is a book I have found useful in general teaching. Different teachers will have their own preferences.

Hartl, D.L. (1994) *Genetics*, 3rd edn. Jones and Bartlett, Boston.

General animal breeding books

Some of these general animal breeding books were published a long time ago, but they represent classic books in their time; some are quite new.

Hill, W.G. & Mackay, T.F.C. (eds) (1989) *Evolution and Animal Breeding*. CAB International, Wallingford.
Hutt, F.B. & Rasmusen, B.A. (1982) *Animal Genetics*, 2nd edn. John Wiley & Sons, New York.
Lerner, I.M. (1958) *The Genetic Basis of Selection*. John Wiley & Sons, New York.
Lush, J.L. (1945) *Animal Breeding Plans*, 3rd edn. Iowa State College Press, Ames.

Mrode, R.A. (1996) *Linear Models for the Prediction of Animal Breeding Values*. CAB International, Wallingford.

Van Vleck, L.D., Pollak, E.J. & Oltenacu, E.A.B. (1987) *Genetics for the Animal Sciences*. W.H. Freeman, New York.

Weiner, G. (1994) *Animal Breeding*. CTA Macmillan, London.

Species books

Cat

Robinson, R. (1989) *Genetics for Cat Breeders*, 3rd edn. Pergamon Press, Oxford.

Cattle

Hinks, C.J.M. (1983) *Breeding Dairy Cattle*. Farming Press, Ipswich.

Schmidt, G.H. & Van Vleck, L.D. (1988) *Principles of Dairy Science*, 2nd edn. W.H. Freeman, San Francisco.

Simm, G. (1998) *Genetic Improvement of Cattle and Sheep*. Farming Press, Ipswich.

Dog

Hutt, F.B. (1979) *Genetics for Dog Breeders*. W.H. Freeman, San Francisco.

Willis, M.B. (1992) *The German Shepherd Dog: A Genetic History*. Witherbys, London.

Horse

Bowling, A.T. (1996) *Horse Breeding*. CAB International, Wallingford.

Pig

Hollis, G.R. (ed.) (1993) *Growth of the Pig*. CAB International, Wallingford.

Sheep

Land, R.B. & Robinson, D.W. (1984) *The Genetics of Reproduction in Sheep*. Butterworths, London.

Piper, L. & Ruvinsky, A. (eds) (1997) *The Genetics of Sheep*. CAB International, Wallingford.

Wickham, G.A. & McDonald, M.F. (eds) (1982) *Sheep Production: Vol 1. Breeding and Reproduction*. New Zealand Institute of Agricultural Science, Wellington.

Journals

Numerous scientific journals in various languages are published throughout the world and many of them have information of value to animal breeders. It is not feasible to list all such journals, but a sample of useful ones are given below. Country of origin and number of issues per year are given in parentheses. Papers do not necessarily stem solely from the country of origin. Journals marked with an asterisk may include papers in languages other than English.

Acta Agriculturae Scandinavica (Scandinavia–4).

Animal Breeding Abstracts (Britain–12). This journal abstracts numerous papers of a genetic nature concerned with animals. It abstracts from over 1000 journals published in numerous languages. It also publishes (at intervals) very useful review articles. It is essential reading for animal breeders hoping to keeping abreast of their field.

Animal Science (Britain–6).

Australian Journal of Experimental Agriculture and Animal Husbandry (Australia–6).

*Canadian Journal of Animal Science** (Canada–6).

Journal of Animal Science (USA–12).

Journal of Dairy Science (USA–12).

*Livestock Production Science** (Europe–12).

New Zealand Journal of Agricultural Research (New Zealand–4).

*World Review of Animal Production** (United Nations–4).

Index